Man's greatest structural achievements!

Here are more than 50 of the world's most extraordinary palaces and pagodas, tombs and towers, monuments and dams. There are canals, temples, churches, and fortresses, too, all superbly illustrated.

The structures range from the ancient to the modern, from the Pyramids, Stonehenge, and the Great Wall of China, to the Astrodome, the Pentagon, and the Geodesic Dome.

A book to amaze, to intrigue, to absorb—for the armchair globe-trotter, the seasoned traveler, and the youngster.

Incredible Structures

JAMES MEYERS

Illustrated by
LUIS DOMINGUEZ

HART PUBLISHING COMPANY, INC.
NEW YORK CITY

COPYRIGHT © 1975 HART PUBLISHING COMPANY, INC.
NEW YORK, N.Y. 10012
LIBRARY OF CONGRESS CATALOG CARD NO. 74-30801
ISBN NO. 08055-1145-8 (Paperback 08055-0167-3)

NO PART OF THIS BOOK MAY BE REPRODUCED OR USED
IN ANY FORM WITHOUT THE EXPRESS PERMISSION OF
THE PUBLISHER IN WRITING.

MANUFACTURED IN THE UNITED STATES OF AMERICA

CONTENTS

Angkor Wat is the world's largest religious structure	10
The Kremlin is the largest fortress in Europe	14
The Pentagon—the world's largest office building	18
The Temple at Madura is ornamented with 30 million idols	22
The Crystal Palace was over a third of a mile long	25
Chichen Itza and the well of death	30
The Winchester Mansion—most bizarre house ever built	34
The Great Pyramid of Cheops—the largest tomb ever built	37
The Great Spinx at Giza is almost 5,000 years old	41
The Panama Canal connects two oceans	43
The Mosque of Omar was built to house a rock	48
The most disastrous structure of this century	51

INCREDIBLE STRUCTURES

The mysterious stone faces of Eastern Island	55
The Potala—the palace that is two miles high	59
The Hoover Dam is over 70 stories high	62
The telltale dome of Florence	66
The towers of the Shwe Dagon Palace are completely covered with gold	68
The Astrodome—space-age arena	71
The longest man-made object in the world	75
The Church of Monte Cassino accommodates only three people	78
The Temple of Borobudar depicts the life of Buddha in sculpture	80
Mount Rushmore contains the world's largest sculptures	84
The Koutoubiya—the perfumed minaret of Marrakesh	87
The Ziggurat—emblem of a great ancient culture	90
The Vertical Assembly Building is the most capacious structure ever built	94
The Tombs of Petra are cut out of the sides of cliffs	97

INCREDIBLE STRUCTURES

The Gateway Arch is the world's tallest monument	100
The most elaborate clock ever built	103
The first highways in America, and the Lost City of the Incas	106
The house built on top of a waterfall	110
The 400 Buddhist temples of Bangkok	113
The smallest drawbridge in the world	116
The Statue of Liberty, a monument to freedom	118
The Purandhar Gate is built on a foundation of solid gold	122
The Chesapeake Bay Bridge-Tunnel—longest deep water bridge in the world	125
The colossal sculptures of Abu Simbel have stood for 32 centuries	129
El Oued—the city of a thousand domes	132
The Leaning Tower of Pisa—the beautiful imperfection	135
The Shayad Tower—emblem of modern Iran	138
The Colossi of Memnon—the statues that sang	141
The fabulous stupas of Anuradhapura	144

INCREDIBLE STRUCTURES

The Verrazano-Narrows Bridge is the longest span in the world	147
The Human Pillar of Oslo	151
The Geodesic Dome becomes structurally stronger as it increases in size	154
Building toward the sky—the story of the skyscraper	157
Mont.-St.-Michel—a village in the sea	162
Stonehenge—an age-old enigma	166
The Alhambra—the fairy-tale fortress	170
The Simplon Tunnel is over 12 miles long	174
The Eiffel Tower is the tallest structure in Europe	176
The Great Wall of China took 1,700 years to complete	180
St. Peter's is the largest Christian church in the world	184
The Taj Mahal—the most beautiful building in the world	189

Incredible Structures

Angkor Wat is the world's largest religious structure

INCREDIBLE STRUCTURES

From the 9th century to the 14th, the city of Angkor was the capital of the Khmer Empire in present-day Cambodia. The home of close to 2 million people, Angkor stretched over 40 square miles in the midst of a jungle. Prosperous Khmer kings constructed immense temples, palaces, and monuments, until Angkor was the largest and most magnificent city in all the Orient.

Sometime in the 14th century, the Khmers were routed and their capital sacked. For the next five centuries, Angkor was deserted. Dense jungle

INCREDIBLE STRUCTURES

growth covered the streets and buildings. Monkeys, bats, and panthers roamed the empty halls. To all the world, Angkor was no more than a legend.

Then, in 1861, Henri Mouhot, a French naturalist, accidentally discovered the jungle city while searching for butterflies. And to the world's amazement, almost every stone in the lost capital was still in place! Travelers and scholars rushed to Angkor to view one of the most marvelous constructions in history — Angkor Wat.

The main temple at Angkor — known as Angkor Wat — is the largest building in all Asia, and the most gigantic religious structure in the world. The temple's vast network of galleries, colonnades, courts, and stairways is crowned by five acorn-shaped towers, the tallest over 250 feet high. Elaborate stone carvings cover thousands of feet of wall space throughout the enormous temple. Images of gods and goddesses, cobras, kings, and Khmer dancers line the rooms and halls, as intricately carved as the most delicate cameo.

Angkor Wat is approached by a 1,200-foot stone causeway passing over the moat which surrounds the temple. Each side of the causeway is bordered by 54 stone genii — eight feet tall — who support the body of a seven-headed stone cobra, the divine serpent of the Khmers. At the end of the causeway, a 65-foot entrance tower leads into the temple.

The central temple is only one of many grand

structures in this vast jungle city. Libraries, baths, gateways, palaces, and temples are spread over an area so large that a visitor could not hope to see all of Angkor in a week's time. The task of clearing away the jungle that chokes the ancient capital is so great that it has never been completed!

In size alone, Angkor is a breathtaking sight, and its temple is a marvel of construction and artistry. That such a magnificent city could remain lost for so long is truly one of the wonders of the world.

The Kremlin is the largest fortress in Europe

Today, Moscow's Kremlin is synonymous with the government of the Soviet Union. But the Kremlin is also a construction of extraordinary beauty and size. In fact, this age-old complex is the largest fortress in all Europe.

INCREDIBLE STRUCTURES

In medieval Russia, a kremlin was a walled bastion within a city which provided protection for the rulers who resided there, and served as the administrative and religious center of the surrounding district. A kremlin customarily included palaces, churches, barracks, storehouses, and markets, and hence, was a small city in itself.

The kremlin at Moscow, now known simply as the Kremlin, was the seat of the Czarist government until 1712, when the Russian capital was

moved to St. Petersburg (now Leningrad). In 1918, after the Bolshevik revolution, the capital was relocated in Moscow, and the Kremlin became the center of administration for the Soviet Union.

This massive city-within-a-city was built in stages over a period of six centuries. The first stone structures were erected in 1365, and the Czar Ivan III rebuilt the entire complex a century later. Over the years, the Kremlin has many times survived the destruction of Moscow itself. In 1812, during Napoleon's occupation, the Kremlin alone withstood the inferno that burned almost the entire city to the ground.

The Kremlin is situated on a small hill overlooking the Moscow River. The fortress consists of a complex of varied buildings surrounded by a triangular wall one-and-one-quarter miles around. In all, the Kremlin extends over an area of 90 acres.

Many of the structures that comprise the Kremlin are world-famous in themselves. The Palace of Facets, built by Italian architects in the late 15th century, is a charming milk-white palace noted for the diamond-shaped facets that adorn its facades. The Grand Palace, built in the 19th century, is the largest building within the Kremlin, and today houses the Supreme Soviet, the parliament of the Soviet Union.

The Spasskaya Tower, one of the Kremlin's 20 gate towers, is the most famous tower in all Russia.

INCREDIBLE STRUCTURES

Nearby, the 270-foot Ivan the Great Bell Tower—the highest structure in the Kremlin—rises to a golden onion-shaped dome.

The renowned King of Bells, the largest bell in the world, is on display near the Bell Tower. This gigantic instrument, cast in 1733, weighs 216 tons and is over 20 feet high. Twenty-four men were required to swing its clapper. Unfortunately, the bell fell to the ground after only three years of use, and has not been tolled since.

The Kremlin also contains the largest cannon in the world, a gun so huge it has never been fired.

On the eastern side of the Kremlin lies the famed Red Square, the site of the incredibly beautiful cathedral of St. Basil. This ornate church, built in the later 16th century, is remarkable for its multi-colored onion-shaped domes. Another feature of Red Square is the black marble tomb of Lenin.

The Pentagon—the world's largest office building

Not long before America's entry into World War II, General B. B. Sommervell proposed the construction of a building to house all the agencies of the U.S. War Department. While many people viewed

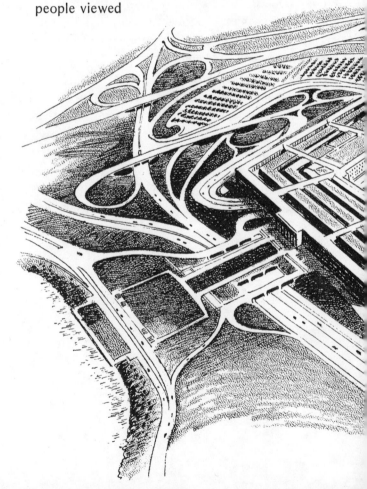

INCREDIBLE STRUCTURES

his proposal as an unnecessary extravagance for a nation that might soon become embroiled in a world war, others felt that this contingency was precisely the reason why the erection of centralized offices was imperative.

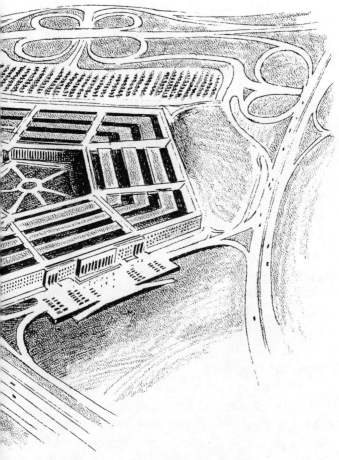

INCREDIBLE STRUCTURES

The latter view prevailed, and construction of the building began in September 1941, on a 34-acre site across the Potomac River from Washington, D.C. More than 13,000 workers were employed on the giant project; 6 million cubic yards of earth were moved; 41,492 pillars were sunk into the marshy earth; 410,000 cubic yards of concrete were poured; and 680,000 tons of sand and gravel were dredged from the bottom of the Potomac. War Department workers began moving into the building even before it was completed in January 1943.

Today, this building—known familiarly as the Pentagon—is synonymous with American military might, and stands as the largest office building in the world. Its total floor area could fill a square whose sides were *one-half mile*. The building consists of five concentric pentagons connected by ten "spokes." The outermost pentagon extends 921 feet on each of its five sides. The innermost pentagon encloses a large open courtyard. Paved courts and roads for delivery vehicles separate the other rings. The ingenious design of the building assures that, despite its size, no two offices are more than 1,800 feet—or six minutes walking time—apart from each other.

The Pentagon is truly a city in itself. Five stories plus a mezzanine and basement comprise a total area of 6½ million square feet, *three times* the floor area of the Empire State Building. Thirty miles of roads and interchanges girdle the site, while 17½ miles of corridors thread through the gigantic complex. Each day, the Pentagon houses more than 30,000 Defense Department workers.

Like any small city, the Pentagon has its Main Street. The long corridor known as the Concourse is lined with shops and showrooms of every kind, from shoeshine parlors and barber shops to airline agencies, a bus depot, and a post office. The two restaurants, six cafeterias, and ten snack bars alone employ a staff of 700.

The Pentagon's total cost of $83 million was considered astronomical at the time of its construction. Yet today, the rental of office space of an equal size would cost the government more than $20 million per day!

The Temple at Madura is ornamented with 30 million idols

According to legend, the Hindu temple of Siva at Madura, India, is adorned with *30 million* intricately molded idols. This figure may indeed be an exaggeration, but even if there are "only" 1 million idols, Madura would still be one of the most elaborately ornamented temples in the world.

In the 16th and 17th centuries, the city of Madura was the capital of a large Hindu kingdom. In this city, King Tirumula began the construction of a massive walled temple complex. But Tirumula's temple was not to be one of delicate grace and beauty to honor a glorious and benign Hindu deity. Instead, the temple at Madura was a shrine to Siva, the Destroyer, and depicted the god in all his grotesque forms.

The walls of this nightmarish temple are over 1,000 feet long on each side and surround a maze of courts, halls, and colonnades. Ten pyramid-like gate towers rise to the height of a 20-story building, and each is completely covered with tier upon tier of densely jumbled idols. These images—molded of plaster, painted in garish hues of red and green, and coated with rancid butter—depict gods and goddesses and demons and monsters of all shapes and forms.

Simply to count all the idols on the towers and walls at Madura would take years. The task of molding and painting them must surely have required decades of work by countless sculptors.

The Crystal Palace was over a third of a mile long

The Great Exhibition of 1851, in London, was the largest and most lavish display of the works of man ever assembled until that time. This international exhibition of machinery, mechanical inventions, manufactured goods, sculpture, and historical artifacts — the forerunner of all subsequent World's Fairs — was housed in an immense glass hall known as the Crystal Palace, which itself was a marvel of architectural and engineering ingenuity.

Surprisingly, the design of the hall was not decided upon until less than a year before the planned opening of the Exhibition. At that time, an open competition was sponsored by the Exhibition's building committee, and within one month 233 plans were received. All were rejected. A plan by members of the building committee was also turned aside. Finally, a plan by Sir Joseph Paxton — submitted in behalf of a private contractor — was accepted by the committee, and work began immediately. Seven months later, the entire Palace was completed.

Paxton's design called for a structure of glass and iron more than one-third of a mile in length, enclosing an area four times that of St. Peter's in Rome. The building was arranged vertically in three tiers, and horizontally in a series of long parallel halls. Raised galleries extending along the interior of the top highest tiers were connected by

pedestrian bridges that passed over the exhibition halls below. The transept, or central entrance hall, was crowned by a vaulted dome 108 feet high, one of the highest glass constructions the world had seen until that time.

The innovative use of glass on the walls and ceiling allowed sunlight to fill the exhibition hall and gave to the interior an "outdoors" effect. Also adding to the natural color was the presence inside the hall of a number of trees that had occupied the Hyde Park and had been left standing.

The Crystal Palace was one of the first structures in the world to demonstrate the exclusive use of iron and glass to enclose a large area. In all, over

4,000 tons of iron were used in the hall, and close to a million square feet of glass — but not one stone added to the support of the structure.

The immensity of the building and the novelty of its structural design led many people to doubt that it could stand up under the stress of the expected crowds. Therefore, numerous experiments were devised to test the strength of the iron framework: three hundred workers were crowded onto one of the galleries and instructed to jump up and down for several minutes; 252 cannon balls were dragged simultaneously over the floor of the hall; an entire corps of military engineers marched several times along the length of a gallery. Yet none of these tests disclosed any weaknesses in the iron girders that framed the hall.

In 1854, with the Exhibition a thing of the past, the Crystal Palace was dismantled and reassembled at Sydenham, England. There it housed a museum and concert hall until the building was damaged by fire in 1936. Three years later, with England at war, the Palace was dismantled completely when it was determined that the structure was serving as a guide to enemy air raiders.

The Crystal Palace had much influence on the structural techniques adopted for subsequent iron and glass buildings. An exhibition hall modeled after the Palace was built in New York City in 1852, for America's first World's Fair. Also called the Crystal Palace, the hall was the first large iron and glass structure built in this country. Reputedly fireproof, the building nevertheless caught fire in 1858 and collapsed into rubble within 15 minutes! Despite this setback to the proponents of iron and glass structures, the influence of the Crystal Palace can be seen today in the glass and steel skyscrapers of the modern city.

Chichen Itza and the well of death

In the centuries preceding Columbus's voyage to the New World, the Maya Indians developed a highly advanced civilization centering in the Yucatan Peninsula of Mexico. One of their greatest cities was Chichen Itza, "the city of the Itzas at the mouth of the wells." First settled by the Mayas in 514, this city was once the home of close to 100,000 people, and the mecca of Mayan pilgrims throughout Central America.

Chichen Itza was abandoned in the 15th century and subsequently overgrown with jungle. But from these ruins modern man has been able to learn much about not only Mayan architecture, but the customs of these ancient people as well.

While all wooden and earthen structures in Chichen Itza had disintegrated long before this century, the stone buildings remain almost intact. These massive white-stone constructions are ornamented with heavy decorative sculpture—with the feathered serpent as the most prominent motif—and enclose dark, cramped interiors.

Among the most striking structures of the ancient city is the Caracol, a round stone tower 41 feet high which probably served as an astronomical observatory, for the Mayas were skilled astronomers. A large stadium and game courts are located near the center of the city. Nearby is the Temple of a Thousand Columns, which gets its name from the rows of stone columns that surround the building.

Yet the most interesting by far of all Chichen Itza's monumental structures is the sacrificial complex leading to the deep well from which the city gets its name. The Mayas built their city beside two such wells: one was used to draw water for irrigation and drinking; the second—called the Cenote—was considered sacred, the home of the rain god Yum-Chac. Into the depth of this sacred well the Mayas hurled precious artifacts and, often, human sacrifices.

Near the Temple of a Thousand Columns, a tall step pyramid rises 100 feet in nine tiers, topped with a small stone shrine. In this shrine began the procession that took the young sacrificial victims to the well of the rain god. Ninety steps lead down one side of the pyramid to a quarter-mile stone causeway.

The solemn procession followed this causeway to a small altar at the edge of the 60-foot well. There, the victims were plunged down into a 60-foot ravine to appease the bloodthirsty Yum-Chac. At the bottom of the ravine lay the well, whose murky waters were sixty feet deep and over 150 feet wide.

The walls of the ravine are too steep to have allowed the victims, had they survived the plummet into the well, to escape the clutches of the rain god. However, these victims were usually so laden with jade and metal that it is unlikely any of them survived the 60-foot plunge or the waters below. When the sacred well was dredged earlier in this century, many precious objects were discovered among the bones in the thick silt below the waters.

The Winchester Mansion—most bizarre house ever built

Mrs. Sara Winchester's mansion near San José, California, is without doubt the most bizarre residence ever constructed. This house — which began as a modest-sized dwelling in 1884 — grew year by year into a nonsensical maze of rooms, corridors, and stairways, many of which served no

function whatsoever. The most imaginative amusement-park funhouse could hardly compete with Mrs. Winchester's mansion in its freakishness of design.

This mansion owes its outlandish construction to Mrs. Winchester's odd fear — she was convinced she would die if she stopped adding rooms to her house! The wealthy woman was so certain of her conviction that she kept scores of carpenters, masons, and plumbers busy every day for nearly 38 years.

INCREDIBLE STRUCTURES

Some rooms in the mansion were built and furnished with the elegance of a royal palace, with gold and silver chandeliers, stained-glass windows, inlaid wood floors, and satin-covered walls. Other parts of the house were constructed only so that the eccentric resident could hear the reassuring bang of hammers. Some rooms measured only a few inches wide, and some stairways led nowhere. The mad mansion contained 2,000 doors and 10,000 windows, many of which opened onto blank walls! The eight-story house also boasted three elevators, 48 fireplaces, nine kitchens, and miles of secret passages and hallways.

When Mrs. Winchester died in 1922 at the age of 85, her mansion contained 160 rooms and sprawled over six acres. The total cost of this insane structure was over $5,000,000!

The Great Pyramid of Cheops—the largest tomb ever built

The pyramids of Egypt were the first great structures ever built by man. In the 5,000 years that have elapsed since their completion, countless structures of magnificent beauty and astounding size have been built, admired, and reduced to rubble. Yet the pyramids remain, in age and in scale the apex of human construction.

Over a period of centuries, thousands of pyramids of varying sizes were erected in Egypt. Each was built to serve as the impenetrable tomb of a royal Egyptian, who, buried beneath millions of tons of rock, would be assured peace and continuing wealth in his life after death. Most of the pyramids have been destroyed by the ravages of time and man, but the three greatest—the pyramids at Giza—remain.

The pyramid of Cheops (or Khugu) is the largest of the three Giza pyramids, and the largest tomb ever built. The immensity of this structure is almost beyond comprehension. For an idea of its size, visualize a huge square tract of land—as vast as, say, Shea Stadium—piled high with rocks to the height of a 40-story building! Within this pyramid, St. Peter's and Westminster Cathedral could both be tucked away with room to spare.

The amount of work required to construct this

INCREDIBLE STRUCTURES

gargantuan tomb is as staggering as its size. The huge stones of the pyramid—each weighing an average of two-and-one-half tons—were quarried miles away from the Giza site, across the Nile River. Each stone had to be cut to size out of solid limestone, ferried across the Nile, and dragged on sledges up to the 100-foot plateau on which the

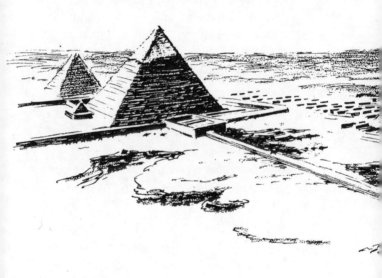

pyramid rests. Then, the stone had to be dragged up a ramp that surrounded the pyramid during construction, and carefully set in place. And this procedure had to be repeated *2½ million times* for the Cheops pyramid alone!

It is estimated that this pyramid required the

work of 100,000 men for a period of 20 to 30 years. In all, 6½ million tons of rock had to be moved to build the pyramid, enough stone to build a wall around all of France. And this work was done without the use of machinery or animals of any kind—not even the wheel!

Despite its immensity, the pyramid constituted only a part of the entire tomb construction project. Dozens of smaller tombs were erected at the base of the pyramid to house the bodies of the king's wives, children, and nobles. A long stone causeway was built connecting the pyramid complex to the Nile. The Great Sphinx was itself a part of the Giza complex. Smaller temples, of both pyramidal and other shapes, were constructed around the larger pyramid, and many of these were massive structures in themselves.

Today, little else but the pyramids remains at Giza. Over the course of thousands of years, the burial chambers deep within the pyramids have been plundered of all their treasures. The white limestone wedges that were used to face the tombs have almost completely disappeared. Yet the pyramids still stand in all their superhuman grandeur, and—alone among all of man's constructions—could very well stand forever.

The Great Sphinx at Giza is almost 5,000 years old

A short distance from the three great pyramids at Giza, Egypt, a curious stone figure of a crouching lion with the head of a man peers solemnly over the ancient Nile Valley. This massive work of carved stone—known as the Great Sphinx—was built almost 5,000 years ago, and is perhaps the oldest monumental statue in existence today.

The sphinx was a mythological beast of ancient Egypt, frequently used to symbolize the Pharaoh in his divine role as the sun god. This beast was usually represented in sculpture by the figure of a crouching lion with the head of a man, ram, or hawk. Thousands of sphinxes were built in Egypt, but the Great Sphinx at Giza is one of the few remaining and by far the largest and most famous.

The Great Sphinx was built in conjunction with the pyramid of Chefren, the second largest pyramid of the Giza complex. A long stone causeway led from this pyramid to a temple in the Nile Valley, near which the Sphinx was erected, facing the Nile and the rising sun. The face carved onto this great stone beast is presumably the face of the Pharaoh Chefren, who wished to remind all those who looked upon his pyramid that it was built to the glory of a divine king.

For centuries, the Sphinx lay covered with

INCREDIBLE STRUCTURES

desert sands, the head alone visible above ground. Age and weather chipped away at the stone head, obliterating the nose and beard and sadly disfiguring the face. But the body lay protected under the sands. When archaeologists recently uncovered the crouching figure, they found the body almost intact.

The entire statue, with the exception of the paws in front, was hewn out of solid rock in one piece. The paws were built of brick. These paws alone could dwarf a man. To appreciate the size of this colossus, imagine a stone figure only slightly shorter in length than a football field, rising almost seven stories off the ground!

And then remember that even this gargantuan figure is dwarfed by those giants of giants, the pyramids.

The Panama Canal connects two oceans

The Panama Canal is neither the longest, the widest, the deepest, nor the oldest canal in the world. Yet, as the only canal which connects two oceans, and the canal whose construction presented the most difficult challenges, the Panama Canal is the greatest man-made waterway in the world.

The initial attempt to build a canal across the narrow isthmus of Panama in Central America resulted in one of the most tragic engineering failures in history. In 1881, a French firm headed by Ferdinand de Lesseps—who had earlier constructed the Suez Canal—began to dig a canal across the isthmus. While de Lesseps was able to conquer the desert of Suez, he could not overcome the mosquito of Panama. Within eight years, close to 20,000 men died of malaria while working on the ill-fated project. The French company went bankrupt after suffering losses totaling $325 million, and de Lesseps left Panama.

In 1907, an American construction crew headed by G. W. Goethals journeyed to Panama to try their luck where the French had failed. Panama leased the U.S. a strip of land 10 miles wide for the canal. A massive project to wipe out the malaria-carrying mosquito was successful, and work pro-

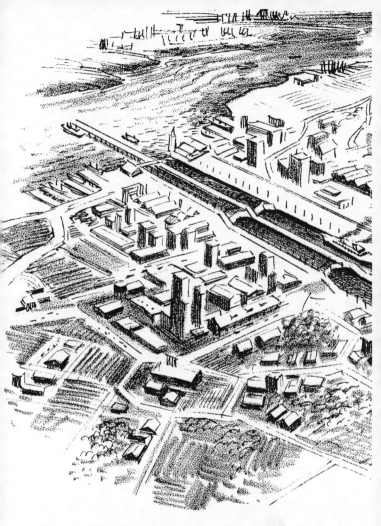

ceeded without the hazard of disease that doomed the French venture.

Construction began at both ends of the projected canal and progressed inland through a dense tropical jungle. An artificial lake was formed; locks

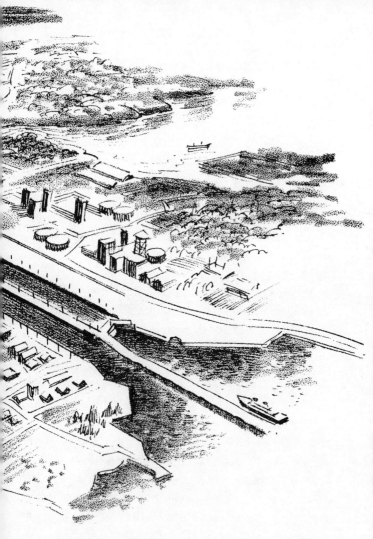

were constructed; the fame Gaillard Cut, for years the largest ditch in the world, was dug through 660-foot Gold Hill. At the peak of construction, 300 railroad cars—each carrying 400 tons of dirt—left the site daily. A total of 10 billion tons of earth were moved for the canal—a figure greater than the total weight of the Great Pyramid of Cheops! In 1914, the canal was completed.

To navigate the canal, a ship must rise or fall a total of 85 feet. To accomplish this, massive locks were constructed which raise a westbound ship first to the level of the artificial lake—Lake Gatun—then to the level of a second lake—Miraflores—and, at the Pacific end of the canal, lower the ship back to sea level. For eastbound ships, the procedure is reversed.

The locks at Miraflores comprise the largest lock system in the world. The largest Miraflores lock is over 1,000 feet long, with gates seven feet thick and as high as a seven-story building. Electric cars, or "mules," on the side of the canal pull the ship through the locks, and water empties in and out of the lock through tubes as large as railroad tunnels.

The St. Lawrence Seaway between Canada and the U.S. extends over 2,300 miles from Duluth, Minnesota, to Belle Isle, Quebec, and hence is more than 45 times as long as the Panama Canal. But the St. Lawrence waterway was built by deepening and widening rivers, and utilizes the Great Lakes for much of its length. The Panama Canal, on the other hand, was carved out of virgin jungle. In addition, the time saved by a seven-hour trip through the canal is a saving unmatched by any other canal in the world, with the possible exception of the Suez (a 9,000-mile journey around South America requires at least two weeks). Each year, over 15,000 ships take advantage of the Panama shortcut.

An interesting footnote to the Panama Canal story: due to the curve of the isthmus, the Pacific end of the canal actually lies 27 miles *east* of the Atlantic end! Also, the canal has provided the world with perhaps its most clever palindrome (a line that can be read in either direction):

A man, a plan, a canal—Panama!

The Mosque of Omar was built to house a rock

The Mosque of Omar in Jerusalem was the first mosque erected with a dome. But this holy structure is more noted for the object it was built to house — a large granite rock sacred to the followers of Islam. Today, the mosque is best known as the Dome of the Rock.

Christians as well as Moslems hold this rock sacred. It is upon this massive stone that — according to Judeo-Christian tradition — Abraham agreed to sacrifice his son Isaac as proof of his obedience to God. In the 10th century B.C., the Hebrew King Solomon constructed his great temple on this site, with the most sacred area of the temple — the Ark of the Torah — directly over the rock. Solomon's temple was subsequently destroyed by the Babylonians, but another temple was constructed on the site by King Herod. This temple was destroyed by the Romans in 70 A.D.

According to Moslem belief, the prophet Mohammed was carried from Mecca to Jerusalem on a winged horse, and set down atop this long-revered rock. Shortly after the Moslems captured the city of Jerusalem in 637, they began to construct a mosque to enclose the sacred stone, preserving it for the followers of Islam.

This mosque, an eight-sided structure topped by a wooden dome almost 100 feet high, was completed in the year 691 and named for the caliph Omar. The courtyard surrounding the mosque occupies almost one-sixth of the entire area of the old walled city.

The rock itself—a massive chunk of granite 60 feet long and 40 feet wide—contains a number of deep hollows, which Moslems hold to be the footprints of the giant horse that brought Mohammed to the sacred city.

The most disastrous structure of this century

In 1940, officials gathered to cut the ribbon on a newly constructed suspension bridge across the Narrows Strait near Tacoma, Washington. The first vehicles routinely crossed the span, and the 2,800-foot Tacoma Narrows Bridge took its place as the third longest suspension bridge in the world.

By this time, the Golden Gate and George Washington bridges had already been in use for several years, and the principles of suspension-bridge construction were thought to be fully understood. Within weeks of the opening of the Tacoma Bridge, however, it was plain to see that more study was called for.

In high wind conditions, the Tacoma span swayed back and forth between its towers. Sometimes the span buckled, forming undulating hills and valleys in the concrete roadway. Cars used the bridge in decreasing numbers as engineers hastened to discover what was wrong with the sparkling-new structure.

Four months after the completion of the bridge, the roadway blew apart in a 42-mile-per-hour wind and tumbled into the waters below. All the theories of suspension-bridge construction were promptly thrown out the window. The only consolation in the baffling catastrophe was the absence of any cars on the Tacoma Bridge at the time of its collapse.

One year before the opening of the Tacoma span, an almost identical bridge had been built in New York City — the Bronx Whitestone Bridge. In view of the Tacoma calamity, engineers began to fear that this bridge might be less than safe.

INCREDIBLE STRUCTURES

INCREDIBLE STRUCTURES

Workers in the then fledgling science of aerodynamics ultimately proved that the designers of the Tacoma span had failed to take sufficient account of the effects of high winds on a suspension bridge. Adjustments were quickly made in the Bronx-Whitestone span, which continues in sturdy operation today. In 1952, a second bridge across the Narrows in Tacoma was erected on the still-standing piers of the first bridge. This second bridge stands today as the eighth longest suspension bridge in the world.

The mysterious stone faces of Easter Island

On Easter Day, 1722, the Dutch explorer Jakob Roggeven chanced upon a small, remote island in the South Pacific, almost 2,500 miles from the coast of South America. There he found — amid the craters of extinct volcanoes and a small tribe of stone-age people — a collection of mysterious faces gazing stoically towards the ocean. Even today archaeologists have not solved all the puzzles posed by the immense monoliths of Easter Island.

The massive heads are set against the gentle slopes of the island's volcanic ridges. Imbedded deep in the soil, the almost identical heads rise from 10 to 40 feet above the ground, and many are estimated to weigh close to 50 tons! In all, over 600 statues dot the island, forming a strange gallery of somber faces on many of the island's slopes.

The figures were carved of tufa, a soft volcanic stone that was quarried in the center of one of the island's volcanoes — Rano Raraku. When explorers discovered the deserted quarry, they found close to 150 additional statues that had never been moved to their intended places on the hillsides. These figures, in various stages of completion, and the tools that still lay scattered about the quarry, gave evidence that the work on the monoliths had been

interrupted quite suddenly and had never been resumed.

The more archaeologists investigated the island, the more they were startled by their discoveries. Bones and ash were found buried in the earth at the foot of the statues. Flat red rocks that lay beside many of the monoliths were shown to be "hats" or "topknots" that at one time rested on top of the heads. And when archaeologists began digging deeper in the soil around the faces, they discovered that the unknown sculptors had carved not merely faces but also full-bodied figures, many of which were now imbedded 30 feet in the ground!

How were the massive stones carried distances of up to 10 miles from the Rano Raraku quarry without losing their smoothly polished finish? How were the gigantic "topknots" hoisted atop the heads without pulleys? How did the bodies come to be so deeply imbedded in the earth? How long ago were the statues carved, and by whom, and for what reason? Why was the work halted so abruptly? All these are questions that decades of research and debate have not answered definitively. It is likely that the puzzles of Easter Island will remain unsolved for all eternity.

The Potala—the palace that is two miles high

Well into this century, Lhasa, the capital of Tibet, was known as the "Forbidden City." In addition to being sealed off from the world by the Himalaya Range and being almost inaccessible by any means of travel, the holy city of Lhasa was closed to foreigners by Tibetan law. For Lhasa has been the home of the Dalai Lama, the spiritual leader of the Tibetan Buddhists. Until recently, few Westerners could claim to have seen Lhasa, or the magnificent palace that overlooks the city: the Potala.

Begun in the year 700, the Potala has served through the centuries as the palace of the Dalai Lama, the seat of the Tibetan government, a college, a monastery, and a fort. Lhasa itself is 12,000 feet above sea level, an altitude at which even the hardiest foreigner will find physical activity extremely taxing; yet the Potala was built high above the city, straddling a steep hill just outside Lhasa. Long zig-zagging stairways provide the only access to the fortress.

The Potala — the "Palace of the Gods" — is a massive complex of buildings which extends for 1,000 feet across the side of the hill and rises nine stories over its lofty foundation. More than 1,400 windows look down over the city. Inside, over 1,000 monks can live and study in the Potala's 500

rooms. The quarters of the Dalai Lama are in a smaller palace within the Potala. For the most part, the fortress is whitewashed and unornamented, built to withstand the bitter Tibetan winter — but the central portion of the Potala is painted red, and its roof and towers are covered with glittering gold!

Since the Chinese invasion of Tibet in 1959, the Dalai Lama has lived in exile in India. The Potala has been shelled by Chinese troops, and many

monks have fled to neighboring Himalayan countries. But in the minds of all Tibetans, the Potala, the palace that soars to the clouds, remains the most sacred building in the world.

The Hoover Dam is over 70 stories high

In the northwestern corner of Arizona, a road emerges from a bleak, dusty landscape and passes over a concrete wall stretched across a dizzyingly deep canyon. On one side of the wall, the Colorado River twists southward through the barren hills; on the other side, a blue lake extends far back along the Arizona-Nevada border. This lake is Lake Mead, formed by the massive concrete wall of the Hoover Dam—the first of the giant concrete dams.

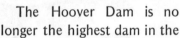

The Hoover Dam is no longer the highest dam in the world. During the last 17 years, eleven higher dams have been built or are under construction. But the Hoover Dam, begun in 1931, pioneered the techniques of modern dam building and remains one of the true structural wonders of the world.

Before the construction of this concrete colossus, the basin now filled by Lake Mead was a canyon cut through the desert plateau by the Colorado River. More than 500 square miles of this

canyon were cleared in anticipation of the waters that would soon begin to flood it. Upriver from the Hoover Dam site, a temporary earthen dam was erected to hold back the river, and four tunnels 50 feet in diameter were cut through the earth to channel the water around the construction site.

The concrete had to be poured without interruption for two years, workers toiling day and night. In 1936, the concrete wall was finally completed, and the four tunnels were sealed. Water began to pile up behind the dam, eventually filling in the canyon and forming the 115-mile long Lake Mead—which until 1955 was the world's largest man-made lake. At the dam, this lake is over 600 feet deep!

To hold back such a massive weight of water, the Hoover Dam had to be built to a thickness of 600 feet at the base— the length of two football fields—tapering to 45 feet at the top. The dam stretches 1,300 feet from one rim of the canyon to the other, and rises 726 feet above the floor of the gorge. In all, over 7,000,000 tons of concrete were used—a weight greater than the Great Pyramid of Egypt.

The benefits of Hoover Dam? A hydroelectric plant at the base of the dam has a capacity of over 1.3 million kilowatts of power, enough to light a good-sized city. And water from Lake Mead is used to irrigate over 1 million acres of land in the U.S. and Mexico.

INCREDIBLE STRUCTURES

For a time, the name of the dam was changed to Boulder Dam, but reverted to Hoover Dam in 1947. Today the dam and lake not only provide electricity, water, and flood control for millions of people, but they also comprise a National Recreation Area visited by thousands every year.

The telltale dome of Florence

The Cathedral of Florence (Santa Maria del Fiore) features one of the most celebrated domes in architectural history. This octagonal ribbed roof structure was designed in 1420 by the first great architect of the Italian Renaissance, Filippo Brunelleschi — with one interesting peculiarity.

The builder left a small opening in the dome, just large enough to permit the passage of a thin

beam of light. A small metal plate was set in the floor of the sanctuary, under the dome. According to the architect's design, the beam of light passing through the opening in the dome should fall exactly on that metal plate each June 21.

This seeming frivolity is not without a purpose, however. Should the beam of light fail to strike the metal plate on June 21, this would constitute a warning that the cathedral had shifted on its foundation. Most likely Brunelleschi himself feared such a shift, for the cathedral was built on a marshy site. Wheeled traffic in the area is prohibited, for fear of disturbing the historic structure.

The towers of the Shwe Dagon Palace are completely covered with gold

Each year, thousands of Burmese pilgrims journey to the capital city of Rangoon to visit the country's most sacred temple, the Shwe Dagon Pagoda. This 15th-century temple-monastery, the center of all Burmese religious life, is most notable for its tall, cone-shaped towers, which are completely covered with gold.

INCREDIBLE STRUCTURES

Shwe Dagon is a complex of temples, reliquaries, towers, and gates, abounding in richly carved ornamentation. Surrounding the complex is a multi-colored tiled terrace 1,420 feet around. Towering above the terrace and gateways are the brick stupas, or sacred relic chambers, each of which is surmounted by a cone-shaped tower covered with gold leaf. The largest of these towers is 326 feet above the terrace, and its glittering gold pinnacle can be seen from virtually any part of Rangoon.

Through the ages, while Western churches have been pillaged, the Pyramids looted, and the Taj Mahal stripped of its gems, here in the center of a large city a fortune in gold has remained untouched. Neither an earthquake that ravaged the city, nor floods, nor heavy bombardment during the Second World War could destroy the fabulous golden cones of Shwe Dagon.

The Astrodome—space-age arena

Upon its completion in 1965, the Harris County Sports Stadium in Houston, Texas — better known as the Astrodome — could lay claim to a number of impressive superlatives: the largest indoor arena in the world, roofed by the largest dome ever constructed, and the only sports stadium in the world that did not contain a single blade of grass.

The Astrodome is truly a landmark of modern engineering and a harbinger of things to come. The stadium sprawls over nine-and-one-half acres and can seat 66,000 spectators. The surface of the field is covered with "Astroturf," a synthetic grass specially designed for the stadium. A scoreboard 474 feet long and four stories high features a 10,000-light screen that can show messages, shorts, cartoons, and anything else that can be put on film. But without doubt, the most revolutionary feature of the $20 million structure is the massive dome.

Constructed of transparent plastic panels supported by a steel lattice, the dome measures 710 feet in diameter and rises to the height of 208 feet over the playing field. The dome of the Pantheon in Rome — the largest dome of the ancient world — had only *4 percent* of the surface area of the Houston dome.

The dome also provides for year-round fair

weather inside the stadium. An air-conditioning system circulates 6,600 tons of air *each minute*, and keeps the temperature inside the stadium at a constant 74 degrees. If on a humid day the air

conditioning were turned off, an entrance of warmer air could cause it to rain in the stadium!

The mammoth dome has resulted in some unexpected problems for players in the stadium. During the first few baseball games played on the new field, players complained that balls hit high into the air could not be seen against the backdrop of the dome. After a number of embarrassing muffs of fly balls by hometown players, the plastic panels of the dome were painted over to provide for better vision.

And, during a baseball game in the 1974 season, Philadelphia's Mike Schmidt did what had been considered impossible—he hit a ball against a public-address speaker affixed to the roof, 329 feet from the plate and 117 feet above the field. The would-be tape-measure however dropped in center field and Schmidt was held to a single.

In 1971, a domed stadium with a retractable panel overhead was completed in Irving, Texas, and the construction of a massive "Superdome" in New Orleans was begun shortly thereafter. Undoubtedly, most of the sports stadiums built in the future will be domed — following the footsteps of the original, the Houston Astrodome.

The longest man-made object in the world

While the Great Wall of China can lay claim to many superlatives, it is not the longest structure in the world. That title goes to a much more modern construction — the fuel pipeline. Because at most points it is buried two feet below ground, a pipeline is certainly not as impressive a sight as the Great Wall, but as an engineering feat it is far more important to our civilization than the Wall was to the ancient Chinese.

Exclusive of highways and railroads, no man-made object can compare with the great pipelines in length. These steel tubes, usually one to two feet in diameter, are most commonly used to carry oil or natural gas from outlying wells to the refineries near big cities.

The longest oil pipeline in the world — owned by the Interprovincial Pipe Line Company of Canada — extends 1,775 miles from wells in Alberta to refineries in Sarnia, Ontario. Along its route, 13 pumping stations move 7 million gallons of oil daily, pumping the oil from three to six miles an hour through the subterranean tubes. However, the world's largest oil pipeline will soon be the Trans-Siberian oil pipeline in the Soviet Union; when it is completed, it will stretch almost 2,860 miles.

An already completed natural gas pipeline between Texas and New York City stretches 1,840 miles, and an incomplete Russian pipeline will extend 2,100 miles. On the drawing boards now are a gas pipeline across Canada with a planned length of 2,200 miles, and a Russian pipeline that would stretch close to 5,700 miles — more than three times the length of the Great Wall!

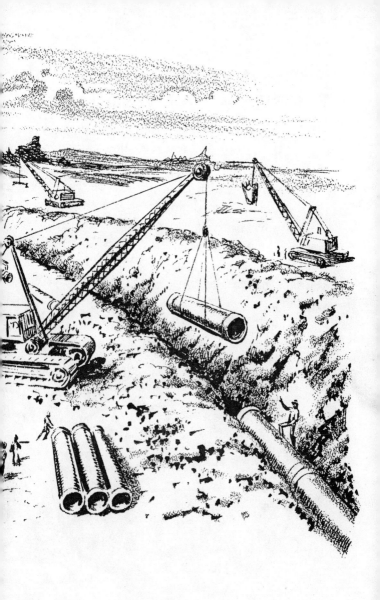

The Church of Monte Cassino accommodates only three people

The church of Monte Cassino, near the city of Covington, Kentucky, is the smallest church in the world.

The church—which from the outside resembles a mausoleum—can accommodate only three people at one time, providing each with a rough wooden bench. The benches face a tiny altar. The walls of the church are only eight feet high, and the church's midget belfry is so tiny that it cannot contain a bell.

The miniature church was built in 1850 by

INCREDIBLE STRUCTURES

Benedictine monks. They named it after the first Benedictine monastery, Monte Cassino in Italy—which was reduced to rubble by bombs in 1944.

The Temple of Borobudar depicts the life of Buddha in sculpture

The Temple of Siva at Borobudar, Java, is one of the largest and, to Western eyes, most peculiar Buddhist shrines in the world. By following a

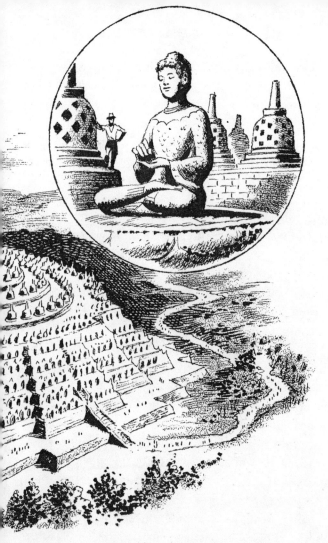

circular route to the top of the pyramidal structure, a visitor can study the life of Buddha and an elaboration of Buddhist religious doctrine—all told in finely carved relief sculpture!

INCREDIBLE STRUCTURES

This reliquary mound was built during the ninth century but lay overgrown with dense jungle for 900 years until its restoration early in this century. The monument consists of seven square terraces of gradating size, surrounded by a high wall and surmounted by three circular terraces or platforms. On these platforms rest 72 small bell-shaped stupas, or reliquary housings; each contains a sculptured Buddha visible through the stupa's perforated stonework. At the pinnacle, one large stupa encloses a large, unfinished Buddha.

Although four stairways—one on each side—rise across the terraces and platforms toward the topmost stupa, the builders of the Temple of Borobudar did not intend that pilgrims climb to the top of the shrine so simply. Rather, they planned a circuitous route lined with teeming relief sculpture. A pilgrim began his visit by walking around the high outer walls, examining the story in sculpture of the Buddha and the various moral doctrines of Buddhism. As he rose from terrace to terrace, the pilgrim inspected the pavilions, the statue-crammed niches, the carved waterspouts, and the relief sculpture that filled the terrace walls.

When read in a circular ascent to the shrine, these reliefs demonstrate, in progressive fashion, ever more profound moral doctrines. The high walls of the terraces prevent a pilgrim from viewing a higher level than the one he is on.

The pilgrim's winding route does not stop when he reaches the circular platforms. Only after he has passed the 72 small stupas is he ready to view the great Buddha that crowns this marvelously ingenious structure.

Mount Rushmore contains the world's largest sculptures

The largest sculptures in the world are the four busts of American Presidents carved into the side of Mount Rushmore, in the Black Hills of South Dakota. Blasted out of a granite cliff, the busts represent George Washington, Abraham Lincoln, Thomas Jefferson, and Theodore Roosevelt. The faces average 60 feet from the top of the head to the chin — proportionate to men *465 feet tall!*

Most of the work on these colossal figures was performed by Gutzon Borglum, who had to blast away close to a million tons of granite as well as chisel in the delicate features of the likenesses. He labored on the project from 1927 until his death in 1941 without seeing it completed. His son Lincoln Borglum finished the monuments later in that year.

Today, the sculptures comprise the Mount Rushmore National Memorial and are known as the Shrine of Democracy. These immense figures are visible from as far as 60 miles away!

The Koutoubiya—the perfumed minaret of Marrakesh

A visitor to Marrakesh, Morocco might believe that his imagination is playing tricks on him when he confronts the Koutoubiya minaret. The tower is a delight to the eye, but of course there is nothing illusory about that. What mystifies tourists is that the minaret seems to give off a delightful smell as well!

The sweet odor visitors detect is quite real. The slender Koutoubiya — a perfect representative of the golden age of Moslem architecture — has been famous for more than seven centuries for the fragrance emanating from its walls.

The story of the sweet-smelling minaret begins in 1195. In that year, the Sultan Yakub al Mansur defeated Alphonso VIII, the Castilian king, at the battle of Alarcos in Spain. To thank Mohammed for the victory, Yakub ordered that a mosque be built at his capital, Marrakesh. And into the mortar used in the building of the mosque were mixed some 960 sacks of musk. That fragrance can still be perceived today.

The 220-foot tower affords a magnificent view of Marrakesh. Yet for more than six centuries, only blind muezzins (Moslem priests) were permitted to climb to the top of the tower. For from that height it would have been possible to look into the open courtyards of the harems of Marrakesh — a sight barred to all men!

The Ziggurat—emblem of a great ancient culture

The ziggurat form was in common use for pyramidal temples in the ancient kingdoms of Sumer, Babylonia, and Assyria. At one time, hundreds of these huge structures—often called step pyramids—stood in the various kingdoms of Mesopotamia (now Iraq). The Biblical story of the Tower of Babel relates one attempt to build such a temple, although ziggurats were never as high as the tale suggests.

INCREDIBLE STRUCTURES

INCREDIBLE STRUCTURES

In ancient times, the most widely known of these structures was the Great Ziggurat at Ur. The city of Ur was the capital of the Sumerian culture and was, the Bible says, the home of Abraham. Under the reign of King Ur-Nammu (C. 2060 B.C.), Ur became the most important city in Mesopotamia, populous and wealthy, an important trading center on the Euphrates River. In the middle of the city, Ur-Nammu built a high terraced ziggurat as the city's chief temple.

This imposing structure, towering above the low mud-brick houses of Ur, was built of sun-baked bricks set in a kind of mortar called bitumen and faced with glazed colored bricks. Ur-Nammu's original temple—built in honor of the moon goddess Nanna — rose in three receding tiers, with three wide frontal stairways leading to the top of the first tier.

Despite the angular appearance of the Ziggurat, there was not one straight line in the entire structure. The ancient Sumerians understood the technique of entasis, whereby supposedly straight lines were curved slightly so that a wall or pillar would appear straight when viewed from the ground.

It was the Babylonian king Nebuchadnezzar who made the Ziggurat at Ur the greatest ziggurat of the ancient world. After the Babylonians had captured the city, Nebuchadnezzar ordered that the Ziggurat be rebuilt. Where Ur-Nammu had built a pyramid of three tiers, Nebuchadnezzar built a seven-tiered tower. Each of the steps in the three frontal stairways was relaid bearing the name of this great king. A series of stairs and passages rising from the first tier gave the impression that a staircase wound around the tower in spiral fashion. Indeed, one could ascend to the apex via a spiral route, but a multitude of other paths were possible.

Nebuchadnezzar and a subsequent king, Nabonidus, more than doubled the height of Ur-Nammu's Ziggurat. At 160 feet, it became one of the highest structures in all Mesopotamia. The base itself — 210 feet by 150 feet — was 40 feet high. At the top of the last tier, a couch and table were left for the moon goddess to use on her visits to earth.

The city of Ur was destroyed many times by conquerors and eventually abandoned. For thousands of years, the Great Ziggurat lay crumbling and covered with sand in the midst of a barren desert. During this century, however, the ruins of Ur were unearthed. The sandy rubble that the archaeologists found can hardly suggest the size and majesty of the ziggurat, the greatest structure of a great civilization.

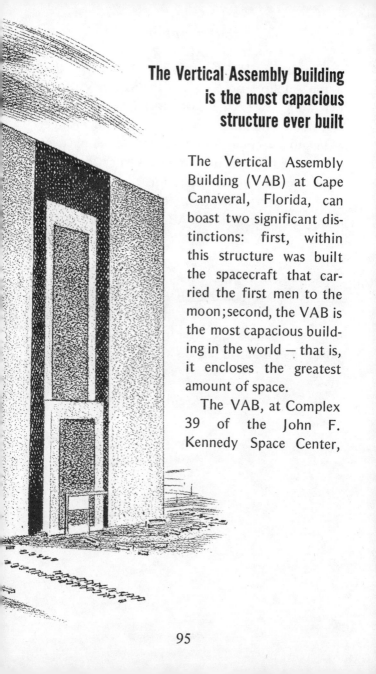

The Vertical Assembly Building is the most capacious structure ever built

The Vertical Assembly Building (VAB) at Cape Canaveral, Florida, can boast two significant distinctions: first, within this structure was built the spacecraft that carried the first men to the moon; second, the VAB is the most capacious building in the world — that is, it encloses the greatest amount of space.

The VAB, at Complex 39 of the John F. Kennedy Space Center,

was completed in 1965 to house the construction of the Apollo spacecraft and the Saturn V rocket that would carry the spacecraft into space. The steel-framed building — 552 feet high and 710 feet in length — could easily contain a number of 363-foot Saturn V rockets in a vertical position. Since the gargantuan Saturn rockets had to be moved from the building to the launching pad upon completion, the VAB can also boast the largest doors ever built—four doors, each *460 feet high!*

The VAB sprawls over 10 acres and has a capacity of 129,000,000 cubic feet. To appreciate the dimensions of this mammoth structure, imagine a hangar the length of two-and-one-half football fields and as high as a 50-story building!

The Tombs of Petra are cut out of the sides of cliffs

The ancient city of Petra—the site of some of the most unusually constructed tombs in the world—lies hidden in a small desert valley deep within the towering red sandstone cliffs of southern Jordan. So remote is this valley that the 2,000-year-old city of Petra remained unknown to the Western world until 1812. When the city was discovered, the whole world marveled at the tremendous pink tombs that were hewn out of the very side of a cliff!

Petra was occupied as early as the 4th century B.C. and was alternately the site of a Roman outpost, a caravan trading center, a Moslem religious enclave, and—during the Crusades—a Christian stronghold. The steep cliffs that surround the valley made Petra an impenetrable fortress, for the valley could be entered only by means of a narrow canyon that winds between the cliffs. This passageway is so narrow that in places a man can stretch his arms and touch both walls! Here, a handful of soldiers could defend the city against an army of any size.

Over the centuries, the rich caravan merchants who made their homes in Petra constructed massive tombs and temples to rival those of any ancient city. Since the narrow entranceway hindered the transport of building materials to the city, the

merchants and Petran kings built their tombs by simply carving out the inner walls of the cliffs.

And what masterpieces these tombs are! Colored rose pink like the sandstone cliffs, adorned with columns and pediments in the classical style, these mammoth structures look as if they were built for giants rather than for men. The largest of the tombs is as tall as a 15-story building, hewn right from the solid rock in one piece! The tomb's main door is over 40 feet in height, and the bottom still is so high that a man would have difficulty climbing it, much less stepping over it. In fact, the tombs are so massive, so incredible in design and construction, that the Arabs who inhabited the surrounding areas believed these tombs could have been built only by jinns, the Moslem spirits.

Throughout the 1,000-foot canyon, these giant tombs and temples are carved from the cliff walls. Each is a marvel in itself, more a colossal work of sculpture than it is a building. Together, the tombs of Petra comprise one of the strangest and most impressive sights in the world.

The Gateway Arch is the world's tallest monument

In the early 1960's, a massive construction project was undertaken in St. Louis, Missouri, to reconstruct much of the city's downtown area — particularly the Mississippi riverfront. The new structures in the restored area included a 51,000-seat sports stadium and the imposing Gateway Arch. Upon its completion in 1965, the Arch became the tallest monument in the world.

INCREDIBLE STRUCTURES

The Arch was built to commemorate the westward expansion of the United States after the Louisiana Purchase of 1803, and especially the

INCREDIBLE STRUCTURES

important role St. Louis played in that expansion as the gateway to the West. The towering, graceful structure — designed by the noted architect Eero Saarinen — spans 630 feet along the Mississippi riverfront. The height of the arch is also 630 feet, which makes the monument as tall as a *60-story building!*

The vivid sight of the sun's rays glistening on the shining steel is visible from miles away. And each year thousands of visitors ride in 40-passenger cars that take them to the top of the colossal arch for a breathtaking view of surrounding Missouri and Illinois.

Though initially the arch met with some opposition from the citizenry at large, it has now come to be regarded as one of the finest achievements in modern architecture.

The most elaborate clock ever built

Visitors to Strasbourg, France will marvel at the gothic splendor of the city's main cathedral, whose 466-foot tower is the tallest medieval tower still standing in Europe. But the modern visitor is likely to be most intrigued by a more recent addition to the age-old cathedral: an astronomic clock that may well be the most elaborate timepiece ever constructed.

The Strasbourg Cathedral, completed in 1439, has displayed an astronomic clock since 1352. But the ingenious device that now stands in the cathedral is the work of Jean-Baptiste Schwilgué, who completed his clock in 1842. This intricate structural masterpiece is run by hundreds of hidden gears, and the various dials and figurines on the outside of the clock form a small mechanical circus, with performances every fifteen minutes.

Near the top of the timepiece, the four ages of man are represented by figures of an infant, adolescent, warrior, and old man. Every quarter hour, each figure in turn takes two steps forward, rings a bell, and then disappears. About the quartet, figures of the 12 apostles appear through a small doorway. A mechanical cock greets the arrival of the apostles by raising its head, bristling its plumage, opening its beak, and crowing three

times. Two angels swing their hammers against bells, adding to the confusion of chimes and moving figures.

Below the angels, seven figures — representing the seven days of the week and the seven inner planets — revolve in a ring, with one figure visible each day. The clock also boasts an angel with a working hourglass, a figure of Christ, a revolving celestial globe, a figure of death who rings a bell with a bone, and sculptured panels depicting Copernicus, Uranis (the goddess of astronomy), the Resurrection, the Creation, and Schwilgué himself.

The clock faces illustrate three distinct methods of time-keeping: standard time, in hours and minutes; apparent time, indicated by: the sunrise and sunset, the phases of the moon, the coming of eclipses, and other movements of the sun and moon; and sidereal time, governed by the movement of the stars through the signs of the zodiac. Together, the various time-telling devices comprise not only one of the most intricate and amusing clocks in the world, but also one of the most accurate.

The first highways in America, and the Lost City of the Incas

INCREDIBLE STRUCTURES

Today's sprawling superhighways may seem to be the ultimate in road construction, but more than 350 years ago, the Inca Indians of South America built a system of roads through the Andes Mountains that would astound even modern engineers.

INCREDIBLE STRUCTURES

At its peak, the Inca Empire extended almost 2,500 miles from Colombia to Chile, and through this mountain country the Incas strung a network of roads to unite their vast domain. As the most important element of the Incas' imperial organization, the road system was built to speed communication and to provide for the efficient movement of men and supplies. The land the Incas ruled was as treacherous as it was vast, and their roads had to cut through marshes, jungles, and soaring mountains. These arteries often climbed heights of over 10,000 feet, and some roads were as long as virtually any modern highway!

Stone retaining walls bordered the roads along much of their length. Enclosed way stations along the route provided for a messenger relay system, and offered protection against sudden mountain storms. Some roads tunneled through mountain cliffs—one such tunnel is almost 750 feet long! At other points, the Incas built causeways over wide marshlands and spanned raging rivers with bridges constructed of twisted rope cables.

The longest of the Inca bridges was immortalized by Thornton Wilder's novel *The Bridge of San Luis Rey*. This 148-foot suspension span crossed a deep ravine of the Apurimac River, swinging precariously in the mountain winds 118 feet above the waters. Until the bridge fell earlier in this century, it was in use longer than any other bridge in America.

INCREDIBLE STRUCTURES

All the roads of the Inca Empire met in their mountain capital of Cuzco, Peru, which is today the oldest continually inhabited city in America. Before moving to Cuzco, however, the center of the Inca empire had been another Andes city, but it wasn't until 1911 that this "lost city" was discovered.

Machu Picchu, 50 miles north of Cuzco, is one of the few urban centers of pre-Columbian America that survives virtually intact. The city straddles a narrow saddle between two peaks, 2,000 feet above a river. Paved stairway streets—stepped because of the slope of the city—weave between stone houses and military fortifications. Hundreds of agricultural terraces give the city the appearance of a gigantic stairway climbing the side of a mountain.

Even today, archaeologists admit the possibility that other Inca roads and cities may still be lost in the Andes, but to this point Machu Picchu is their greatest find.

The house built on top of a waterfall

"Fallingwater" is the popular name for one of the best-known private residences in the country, the Edgar J. Kaufmann house on Bear Run, Pennsylvania. The name is appropriate, for this home — designed by America's premier architect, Frank Lloyd Wright — was built directly over a waterfall!

INCREDIBLE STRUCTURES

When Mr. Kaufmann commissioned Wright to design a weekend retreat for his family on his Pennsylvania property, he wanted the architect to take full advantage of the picturesque site: a wooded glen with a slowly running stream, a clear pool, and a small waterfall. And Wright more than fulfilled his expectations, designing a striking three-story structure anchored on a small cliff overlooking the pool, with a portion of the house cantilevered directly over the running water.

Built almost entirely of masonry, the home features six reinforced-concrete terraces extending over the waterfall and pool. Most rooms offer access to a terrace and a breathtaking vista of the waters below. A stairway suspended from the lower story reaches to within a few feet of the waterfall itself. Viewed from the pool, Fallingwater seems to rise from the boulders around it, and the stream appears to run directly through the house, as if the structure were part of the natural site rather than an addition to it.

The unusual site and construction of Fallingwater earned the home much publicity when it was completed in 1936, and the home remains one of Wright's best-loved works. Fallingwater is open to visitors.

The 400 Buddhist temples of Bangkok

Bangkok, Thailand, is the home of some of the most stunning Buddhist temples in the East. In all, there are some 400 Buddhist monasteries within the limits of this canal-crossed city, often called the "Venice of the East."

Perhaps the most well-known of Bangkok's many religious structures is the Temple of the Emerald Buddha. Since the 15th century, when King Tiloka adopted the structure as the spiritual safeguard of Thailand, the Temple of the Emerald Buddha has been the center of all Thai religious life. The Temple is the home of the most sacred objects in Thailand, chief among these an immense statue of a meditating Buddha mounted on a pedestal under the Temple's high roof.

The Emerald Buddha forms only a part of a larger religious complex, called the Wat Phra Kaeo, situated on the banks of the Chaophraya River in the Thai capital. The complex is surrounded by a wall four and one half miles long, 13 feet high, and 10 feet thick. Sixty-three ornamented gates permit entry to the sacred grounds.

In addition to the Emerald Buddha, the Wat Phra Kaeo includes a depository of ancient Buddhist scriptures, memorials to white elephants, statues of venerated holy men, and a great stupa or reliquary. The Royal Pantheon — which is open to

visitors only one day each year — contains life-sized bronze figures of former Thai kings. Throughout the grounds, tall plaster demons, or Yaks, have been placed to ward off evil spirits.

In every building of the temple complex, each door, window, statue, tower, and pillar tapers upwards. The most striking features of these buildings are the "sky licks," curving, pointed

pieces of ornamental metal that resemble licks of flame.

Directly across the Chaophraya River from Wat Phra Kaeo lies the small Wat Arun, or Temple of the Dawn. Here, as in the Wat Phra Kaeo, the brick walls are intricately inlaid with bits of shell, pottery, and porcelain. The various levels of the Wat Arun are supported by rows of columns sculpted in the form of demons.

The two temple complexes, bordering the river with their needlepoint towers glittering in the sun, form one of the most memorable panoramas in the East — or for that matter, in the world.

The smallest drawbridge in the world

The world's smallest drawbridge is located at Ely's Harbor on the island of Bermuda. This tiny span—known as Somerset Bridge—consists of two masonry embankments connected by two tiny wooden draws.

The entire bridge is less than 20 feet from one bank of the waterway to the other. The two draws in the center are each shorter than the width of the bridge. When the draws are pulled up, the opening in the bridge is barely wide enough to permit the passage of one sailboat!

The Statue of Liberty, a monument to freedom

Towering above the harbor of New York, the gateway to America, the Statue of Liberty has stood as a welcoming beacon to millions of immigrants and visitors for close to 100 years.

INCREDIBLE STRUCTURES

What a thrill it is, after a long voyage, to slip through the Narrows Channel into the Upper Bay and suddenly view the vast panorama of New York City, with the great lady standing tall in the midst of the busy harbor, facing the open sea, her torch held high above her head in welcome.

The full name of this world-famous monument is The Statue of Liberty Enlightening the World. As befitting a monument to universal liberty, the Statue was erected not by any monarch or government, but by citizens of France and the United States, the twin vanguard of democracy.

INCREDIBLE STRUCTURES

An organization known as the Franco-American Union was founded in 1875, the year before America's centennial celebration. A proposal by the Frenchman Edouard Laboulayé was adopted whereby donations from American and French citizens would be used for the design and construction of a monument to commemorate the American and French Revolutions, and to symbolize the long-standing friendship between these two nations.

French citizens raised money for the statue, and a design was accepted from Frederic Auguste Bartholdi for an iron-and-steel frame statue covered with copper sheets. American citizens raised the money for the granite-and-concrete pedestal. Bartholdi, during a visit to the United States, suggested that the statue be placed on 12-acre Bedloe's Island, in the middle of New York Harbor.

American builders constructed the pedestal atop an 11-point star formed by the walls of old Fort Wood, which had previously occupied the island. Rising atop the pedestal was the iron-and-steel frame designed by Charles Eiffel—who was later to build the Eiffel Tower. When the 300 sculptured copper sheets were fixed to the frame, the copper-colored colossus was left towering above the bay. (Copper oxidizes with age, and today the statue is entirely green.)

INCREDIBLE STRUCTURES

The Statue of Liberty was dedicated on October 28, 1886, and in 1924, it became a National Monument. In 1960, the name of Bedloe's Island was officially changed to Liberty Island.

The statue itself is a 152-foot figure of a woman in long robes, raising a lighted torch above her head. In her left hand she holds a tablet on which the date July 4, 1776—the date of the signing of the Declaration of Independence—is inscribed. Around her feet lie broken chains, symbolic of the breaking of the bonds of oppression that the statue commemorates. The pedestal adds nearly another 150 feet, giving the monument a total height of over 300 feet.

Visitors can take an elevator to the top of the pedestal, then climb a narrow, winding staircase that twists up into the Lady's crown. There, an observation room 260 feet above the water affords a spectacular view of New York Harbor. In earlier days, visitors were permitted to climb through the arm of the statue and emerge on a platform surrounding the torch, but this passageway has been closed off in recent years, due to suspected structural weaknesses in the arm.

Whether one views the city from atop the statue or admires the statue from the city and harbor below, the Statue of Liberty remains an inspiring monument, a structural wonder, and the symbol of a nation.

The Purandhar Gate is built on a foundation of solid gold

For sheer extravagance, the Fortress of Purandhar near Poona, India, may be without equal in the world. While the fortress itself is less than luxurious, the gateway to this ancient structure was built upon a foundation of *solid gold!*

The Rajah of Bedar ordered the construction of

the fortress in 1290. However, his engineers soon informed him that the site for the planned gateway was so marshy that any structure built there would soon sink into the ground. According to legend, the Rajah followed the advice he received in a dream and decided to construct the gateway on a foundation of solid gold.

Engineers dug two large cavities in the earth, each 12 feet deep. Into these pits they laid 50,000 14-karat gold bricks, weighing a total of some 27,000 pounds. At current prices, the cost of the foundation alone—which emptied the Rajah's treasury—would be over *$40 million!*

It might be added that the foundation did achieve its aim: the gate of Purandhar still stands today, almost 700 years later!

The Chesapeake Bay Bridge-Tunnel—longest deep water bridge in the world

At the mouth of the Chesapeake Bay there is a narrow causeway which stretches as far as the eye can see. When you cross the bridge and reach the middle of the bay, you are completely out of sight of all land.

When engineers began planning the Chesapeake Bay crossing, they knew that the channel's width — over 17 miles at the narrowest point — called for the construction of a low causeway. Yet through this channel would pass some of the largest ships in the world, bound for Baltimore, Norfolk, and other cities that make up the largest single port area in the United States. These deep-water vessels could not pass under a causeway. A tunnel under the ship channels seemed to be in order, yet the cost of a 17-mile long tunnel—five miles longer than the longest tunnel ever built—was prohibitive. So the engineers decided to build both—a combination causeway and tunnel that would carry vehicles across the bay yet permit ships of any size to pass over the roadway tunnels.

Construction of the Chesapeake Bay Bridge-Tunnel began in 1961. Four man-made islands were built in the bay to anchor two tunnels — each over a mile long — directly under two ship channels. Close to 15 miles of causeway was

constructed on pillars rising 30 feet over the bay, with a 28-foot-wide roadway on top. Higher trestle bridges were built near both shores.

In 1964, after 37 months of work and over $210 million in construction costs, the bridge-tunnel was completed, stretching 17.6 miles between Cape Henry and Cape Charles, both in Virginia. Today, for a toll in excess of $5, drivers can save hundreds of miles of travel around Chesapeake Bay.

INCREDIBLE STRUCTURES

The Chesapeake Bay Bridge-Tunnel is not the longest bridge in the world. That title belongs to the bridge across Lake Pontchartrain, a 24-mile causeway connecting New Orleans and Lewisburg, Louisiana. But the Louisiana bridge crosses a shallow lake. The Chesapeake Bay Bridge-Tunnel traverses a deep ocean bay, and its combination of bridge, tunnel, and causeway makes the Chesapeake, if not the longest, at least the most unusual water crossing in the world.

The colossal sculptures of Abu Simbel have stood for 32 centuries

Ever since the 13th century B.C., men have marveled at the mammoth stone statues outside the temple of Amon Re, hewn out of the side of a steep cliff at Abu Simbel, Egypt. Each of four gigantic sculptures — the largest is close to 70 feet high — depicts the great Pharaoh Ramses II seated on a massive stone chair, facing eastward over the Nile.

During Ramses' 67-year reign, Egypt acquired unprecedented wealth and power, and the Pharoah left monuments throughout his empire. Among these were the temples at Luxor, Karnak, and Abu Simbel, three of the greatest works of ancient Egyptian architecture.

After Ramses' reign, no one could sail up the Nile without coming under the gaze of the four watchful statues on the western bank — and the large figures before another temple slightly to the north.

At the foot of the four seated figures stand a number of smaller statues. One is set directly in the wall behind the colossi. Entrance to the temple is gained by stairs between the two center statues, which lead to a narrow hall passing between two rows of standing figures inside the temple.

In the 1960's, it appeared that these age-old

monuments would be lost forever. The Egyptians were constructing the Aswan High Dam upriver from the temple site, and the water that would flood the Nile Valley for 310 miles behind the dam threatened to submerge many archaeological treasures — among them Abu Simbel.

International negotiations aimed at preserving the statues resulted in the adoption of an Italian plan for moving the entire temple complex to higher ground. Money for the conservation project was allotted by UNESCO, an agency of the UN. An international team of archaeologists and engineers set to work, sawing the statues into massive blocks and reassembling them 200 feet above the original site.

In 1970, the Aswan High Dam was completed. The original Abu Simbel site is now submerged under Lake Nasser. But the statues of Ramses — after 32 centuries at one site — still gaze stoically over the Nile Valley.

El Oued—the city of a thousand domes

The city of El Oued, Algeria, is nestled in a large oasis deep in the Sahara Desert. Surrounded by the Great Eastern Erg—the most barren region of the entire Sahara—El Oued basks in the fierce desert

sun for 365 days a year, almost untouched by rainfall.

The residents of this age-old caravan trading center have constructed their one-story homes with thick mud walls for protection against the scorching sun. To further reduce the heat within these structures, almost every *room* in the city is roofed by its own mud dome. While some homes are topped with a single oblong dome, others sport clusters of small circular domes which give the densely packed city a truly bizarre appearance. From the air, El Oued looks like a bunch of egg cartons turned upside down!

The Leaning Tower of Pisa—the beautiful imperfection

One of the best-known towers in the world owes its fame neither to beauty nor to size, but to imperfection. Hovering on the apparent brink of collapse, the campanile of the Cathedral of Pisa—better known as the Leaning Tower of Pisa—attracts thousands of visitors each year to its precarious galleries.

Apart from its obvious tilt, the tower is an exquisite work of architecture, a free-standing bell tower in the Italian Romanesque style. The tower's cylindrical core is surrounded by six galleries of marble columns, crowned with a multi-colored belfry. Visitors can climb the tower and step out onto the deck encircling the belfry for a dizzying look at the cathedral and streets below.

Construction of the tower began in 1174. A stone-bed foundation was laid 62 feet deep in the marshy ground. Tree trunks were used as piles. Nonetheless, a tilt to the south soon became apparent. To compensate, the builders increased the height of the galleries on the south side of the tower. Further adjustments were made during the 13th century, but the tower only leaned further towards the south.

Actually, not only does the tower lean to the south, but—due to the extra height added to the

south side—it curves towards the north as well. The 179-foot tower is now *14 feet* off perpendicular!

The famed astronomer and physicist Galileo is believed to have put the tower's tilt to good use. As the story comes down to us, in 1589 he conducted an experiment by dropping objects from the upper galleries. The tower's lean assured Galileo that the falling objects would not strike the side of the structure on their way down. These experiments helped the physicist to formulate his laws of motion, which maintain that objects fall at the same speed and with uniform acceleration, regardless of their weight.

Various proposals to save the tower have been put forth through the years. One of these proposals must work if the tower is to survive; otherwise, collapse is only a matter of time.

A similar leaning tower in Saragossa, Spain, fell in 1887. But even if the Tower of Pisa is saved by engineers, one thing is certain: with its southward tilt and its northward curve, the tower will never stand straight.

The Shayad Tower — Emblem of Modern Iran

In 1971, Iran —— which describes itself as the world's oldest monarchy — celebrated the 2,500th anniversary of the first Persian Empire. Heads of

INCREDIBLE STRUCTURES

state from around the world came to view the gala celebration staged near Persepolis by Shah Mohammed Reza Pahlavi, the reigning monarch since 1941. And in Teheran, the capital city, construction began on a striking new monumental

tower, the Shayad, to commemorate the anniversary.

The Shayad Tower was completed in 1972, and stands today as the gateway to bustling Teheran. The bold design of the tower combines both modern and traditional forms, symbolizing the illustrious past and the promising future of this oil-rich nation.

The Colossi of Memnon—the statues that sang

In the 15th century B.C., the Egyptian king Amenhotep III erected a funeral temple near the city of Thebes, with two colossal stone statues guarding the entrance. In the following centuries, Egyptians were startled to hear, at each dawning, mysterious musical sounds emanating from one of the colossi!

The Greeks, equally baffled by the harp-like noises, named the 60-foot statues after the demigod Memnon. The daily song, they believed, was the voice of Memnon greeting his mother Eos, the goddess of dawn.

After an earthquake damaged the two colossi, the Roman Emperor Septimius Severus had the statues repaired. But when the restoration was completed, the strange cries of Memnon ceased forever, as mysteriously as they had begun (although visitors today sometimes claim to hear eerie sounds emanating from the statues). Today the funeral temple is gone and the colossi stand on the desert alone — and silent.

The explanation for the strange cries? The rapid change in temperature as the desert sun rises at dawn produces strong air currents. These currents probably resounded through the loose joints of the colossus before Severus repaired it. The acoustic

INCREDIBLE STRUCTURES

principles responsible for the curious sounds are similar to those of an organ pipe — making the statues the most oddly shaped organ pipes in history!

The fabulous stupas of Anuradhapura

A stupa is a Buddhist monumental mound built to house a sacred relic. Usually constructed of earth

faced with stone, these often massive reliquaries can be found in all the nations in which Buddhism is practiced, and many stupas are over a thousand years old. One of the greatest and oldest collections of stupas in the world is found in Anuradhapura, on the island of Sri Lanka, or Ceylon.

This ancient city served as the capital of Ceylon for 12 centuries and was a center of pilgrimage for many Buddhists. Today, the ruins of Anuradhapura include several colossal stupas (some larger than the great pyramids of Egypt), a temple hewn from solid rock, and the Brazen Palace, so named for its roof of brass.

But the most marvelous of all the sacred structures in Anuradhapura is the Ruwanweli Pagoda. This stupa, built in 144 B.C., is constructed on a base of *solid silver,* over 500 square feet in area and seven inches thick. The value of the metal used in the foundation alone has been estimated at close to 3 million dollars!

The Verrazano-Narrows Bridge is the longest span in the world

For many years, the entrance to New York Harbor through the Narrows Strait was considered too wide to be spanned with a bridge. Due to the depth of the Strait and its heavy ocean-going traffic — all ships entering the port of New York must pass through this channel — the use of a causeway, cantilever, or any other kind of bridge calling for the placing of pillars in the Strait was ruled out.

The construction of the Brooklyn Bridge in 1883 demonstrated that a suspension bridge could span a sizable distance without pillars obstructing the passage underneath. But the Brooklyn Bridge spanned only 486 feet, a distance hardly comparable to the 4,000-foot-plus width of the Narrows. Then, in 1937, came the completion of the Golden Gate Bridge in San Francisco, with its unbroken span of 4,200 feet. This achievement confirmed the capabilities of the suspension bridge, and plans for a bridge across the Narrows were begun in earnest.

A visitor to New York sailing through the Narrows in 1959 would have observed the bridge taking shape on the shores of Brooklyn and Staten Island. In that year, two steel towers began rising near the sites of Fort Hamilton and Fort Wadsworth, the old fortresses that had guarded the entrance to New York Harbor.

Two years later, a visitor would have found the tall gray towers completed, and thick steel cables strung above him across the busy Strait. By early

148

1962, a small segment of the roadway structure was suspended from the cables in the very middle of the Strait — for the construction of the roadway began in the center of the span and proceeded toward the two towers.

In 1964, after five years of work and $325 million in construction costs, the longest span in the world was opened to traffic. The bridge was named the Verrazano-Narrows after the Strait it spanned and the Italian explorer who was the first European to sail into the harbor.

From tower to tower, the span extends 4,260 feet, 60 feet more than the Golden Gate span. The two towers — as tall as 70-story buildings — are so far apart that they were constructed five inches out of parallel to allow for the curvature of the earth! The highest point of the roadway is 228 feet above the water and extends between its two anchorages a distance of 6,690 feet!

One year after the opening of the Verrazano-Narrows Bridge, a second roadway level—which had been almost completely constructed but had not been intended for use for at least another ten years — was quickly put into operation as traffic on the bridge far surpassed all expectations. The two roadways provide for 12 lanes of traffic and weigh more than 60,000 tons!

The Human Pillar of Oslo

Rarely is the work of an artist so closely associated with a city as the work of Gustav Vigeland is with Oslo. Few cities can boast a more impressive outdoor art attraction than the Frogner Sculpture Park, which is devoted entirely to the work of Vigeland, Norway's most famous sculptor.

In 1921, the 51-year-old Vigeland — already a renowned artist — entered into a unique contract with the Oslo municipal authorities. He agreed to bequeath to the city almost all of his work, both past and future; in return, the city built Vigeland a spacious studio where the artist lived and worked until his death in 1943. Today the studio-residence forms the Vigeland Museum, which contains 1,650 sculptures and thousands of Vigeland's woodcuts and sketches.

But Vigeland's greatest gift to Oslo is the nearby Frogner Sculpture Park, which the artist designed to display his most monumental works. The verdant park, over one-half mile long, contains hundreds of figures carved in granite, iron, and bronze. The works include a bridge with 58 bronze figures adorning its parapets, and an ornate fountain.

The most spectacular of all Frogner Park's many sculptures is the obelisk known as the Human Pillar. Vigeland began designing the Pillar in 1924,

INCREDIBLE STRUCTURES

but not until 1943 was the cutting of the stone completed. The Pillar is set in the middle of a circular mount of broad steps, on which 36 large granite groups of figures are set. Cut from a single piece of stone, the Pillar rises 55 feet, with a base diameter of eight feet, and contains 121 figures of men and women at all ages of life clambering in a spiral toward the summit. At the bottom, corpses are piled on top of one another; at the top, children are held aloft by their mothers. The complex intertwining of forms, symbolizing the development of man, displays a superb mastery of modeling and detail.

Frogner Park was opened to the public in 1947, and is visited by thousands each year. The Human Pillar stands as Vigeland's most memorable work, and one of the finest non-functional structures ever created.

The Geodesic Dome becomes structurally stronger as it increases in size

The geodesic dome, designed by the American architect, engineer, and inventor R. Buckminster

Fuller, may turn out to be the most important structural innovation of the 20th century. This dome is built with a frame of interlocking tetrahedronal shapes copied from nature, as in the structure of crystals or plant cells. Fuller's dome is the only structure yet devised whose strength actually *increases* with its size!

Because of this fortunate peculiarity, a geodesic dome could be built to enclose an area of any size. Theoretically, entire cities could be covered with such a dome to form a climatic environment which would be completely controllable by man. Even more intriguing is the possibility of enclosing a currently nonarable expanse of land, such as the Sahara or the Antarctic, and thus rendering it fertile!

The idea of bubble-topped cities may seem to be more in the realm of science fiction than engineering fact, but thousands of geodesic domes have already been constructed throughout the world. To date, the largest is the Union Tank Car Company in Baton Rouge, Louisiana. Completed in 1958, this dome is 384 feet in diameter and 116 feet high. However, many larger domed projects have been proposed, among them a shopping center in East St. Louis, Illinois, that calls for the construction of a geodesic dome close to a *half mile* in diameter!

Building toward the sky—the story of the skyscraper

The skyscraper is the symbol of the modern city. Today, cities of even modest size boast 20- or 30-story buildings, and almost every large city in the country is characterized by the contours of its 30- and 40-story skyline. Yet as late as 90 years ago, a building of 10 or more floors was considered a giant!

Until the 1870's, the need for thick masonry walls and the lack of a sophisticated elevator limited most office buildings to a height of five or six stories. In 1870, the Equitable Life Insurance Company Building in New York broke new ground by rising to a height of 130 feet. But, like earlier structures, the Equitable Building was supported by masonry walls. The first building to employ a steel skeleton for support, and thus the first true skyscraper, was the Home Insurance Company Building, erected in Chicago in 1885.

After the invention of the electric elevator in 1887, the title of the world's tallest building passed in rapid succession to a number of structures. The 612-foot Singer Building in New York claimed the title upon its completion in 1908, but this height was surpassed just a few months later by the 700-foot Metropolitan Life Building, also in New York. In 1913, the Woolworth Building in New

York took over as the world's tallest, at an altitude of 792 feet. Then in 1931, came the first building to rise over 100 stories—102, to be exact; this 1,250-foot tower was New York's Empire State Building.

During the 1940's and 50's, the advent of the curtain-wall building (a building almost entirely walled with steel and glass) brought a new shape to the skyscraper. Office buildings tended to be larger in base area but shorter in height, and it began to seem that the Empire State Building would retain its crown forever. In the late 1960's, however, the plan for a new office complex in New York was announced: the World Trade Center, a project that called for the construction of *two* 110-story towers side by side!

By 1974, the Center was virtually complete on its 16-acre site in Lower Manhattan. The two 1,350-foot towers had been topped off, and many tenants were conducting their business in what had become the tallest buildings in the world. In addition to the two towers, the World Trade Center consists of a new U.S. Customs House, three smaller buildings, a five-acre plaza, and a new terminal for the PATH commuter rail line — the first completely air-conditioned subway station ever constructed.

The two towers are striking in scale and in simplicity of design, and stand out boldly even from the colossal skyline of Manhattan. The steel-and-glass-faced structures are oblong in shape, flat-roofed, and 209 feet square at the base. Although the two towers occupy only slightly more than two acres of ground area, they can accommodate 50,000 workers and 80,000 visitors daily!

Naturally, the designers of such a gigantic complex were confronted with numerous problems, each of which they solved with innovation and daring. For example, more than 1.2 million cubic yards of earth and rock had to be excavated from the site, and the job of transporting such a massive load from Manhattan to a distant land-fill site seemed overwhelming. So, the builders decided to dump the fill at the adjoining Hudson River shoreline, thus creating 23.5 acres of new land to be deeded to the City of New York. Another problem — how to wash 43,600 windows — was solved with automatic window-washing machines that slide up and down the facades of the buildings in stainless-steel tracks.

To provide sufficient elevator service for so large a building would seemingly require that close to half the floor area of each tower be devoted to elevator shafts. The solution to this problem was the "sky lobby." The towers contain lobbies on the 44th and 78th floors as well as the first. Passengers bound for the upper two-thirds of the building board large express elevators which speed them to a sky lobby, where they board a second, local elevator that takes them to their floor. The system assures that no passenger will have to stop at more than six floors before reaching his own. More importantly, since the local elevators run only between one lobby and another, three elevators can actually use the same shaft! Thus, the 104 elevators in each tower take up only 13 percent of the floor area, as compared to the 23 percent required for shafts in most office buildings.

The World Trade Center was not to reign for long as the world's tallest structure. At 1,450 feet, the Sears Tower recently constructed in Chicago tops the New York towers by 100 feet. How long the Chicago giant will be the highest in the world is anyone's guess. Engineers claim that it is now structurally feasible to construct skyscrapers over a mile high!

Mont-St.-Michel—a village in the sea

In the English Channel, one mile off the coast of Normandy, the rocky islet of Mont-St.-Michel rises from the misty sea like a mirage. In this strangely

beautiful village surrounded by medieval walls, ancient houses climb a steep hill towards the towering abbey spire. It is a truly breathtaking sight.

The story of Mont-St.-Michel is as remarkable as its beauty. Hundreds of years ago, the land on which the village rests was part of the mainland of France, a tree-clad granite hill rising 260 feet above an oak forest. A pagan shrine stood at its summit. The English Channel washed the shore, miles away from the hill.

When the Romans invaded northern Europe, they replaced the old shrine with a temple to Jupiter. Centuries later, early Christians erected a crude chapel atop the rocks. In 708, Benedictine monks built an abbey on the site of the old chapel, overlooking a rich plain where animals grazed and farmers toiled in the salty air of the distant sea.

One day in the year 725, an earthquake shook the coast. A tidal wave surged inland, devastating the plain. When the waters receded, the farmlands were no more; the hill stood amid a vast stretch of tidal flats. The coastal village of Mont-St.-Michel had become the islet of Mont-St.-Michel.

The monks who had occupied the abbey atop the crag, were convinced that God had spared their holy village from the ravages of the tidal wave, and they remained on the island. Pilgrims from all over France flocked to the village to visit the blessed abbey, and the town of Mont-St.-Michel became rich and powerful through their donations.

The high stone walls of the village and the natural moat formed by the daily rush of the tides made Mont-St.-Michel an invincible stronghold. When all of northern France had fallen into the hands of the English during the Hundred Years War, Mont-St.-Michel remained French. Many times besieged, the island village was never captured.

Even in the centuries since the earthquake, Mont-St.-Michel has never truly become an island. Twice each day, a 40-foot tide—one of the highest in the world—rushes in over the tidelands and leaves the village surrounded by water. When the tide recedes, the village is left in the midst of marshy wetland, linked to the shore by a half mile of beach. In 1875, a stone causeway was built to provide access from the mainland in high tide.

Today, 250 people live in tightly packed houses on the tiny enclave. The abbey, as it has for over 1,200 years, lifts its spire 500 feet above the waters around it—the waters that could not conquer Mont-St.-Michel.

Stonehenge—an age-old enigma

One of the world's simplest, yet most astounding structures—Stonehenge—lies in eerie solitude on the marshy Salisbury Plain of Southwestern England. This remarkable construction of massive stones, built by an unknown people thousands of years ago, has been a puzzle to archaeologists and historians for centuries.

INCREDIBLE STRUCTURES

Although many of the stones have now fallen to earth, we know that the original arrangement of ditches, holes, and rock constructions was basically a series of concentric rings. A circular ditch 300 feet in diameter forms the outermost ring. Moving in toward the center, the next ring consists of 56 circular holes filled in with earth. These "Aubrey

holes," so-called after the British antiquarian who studied Stonehenge—are each six feet wide and four feet deep. Within this ring are two circles comprised of smaller filled-in-holes, known as the Y and the Z holes.

The third ring is a circle of large Sarsen stones, each about 13½ feet high, arranged in post-and-lintel (upright-and-crossbeam) formations. The innermost ring is a circle of upright Bluestones, without lintels.

Within this Bluestone ring, we find a horseshoe of five hugh trilithons: massive stones as much as 24 feet high, with lintels across the top of the upright post stones. Each trilithon weighs over 30 tons! The horseshoe surrounds an ovoid formation of Bluestones, which is in turn wrapped around the center "altar" stone.

The amount of work required of a prehistoric people just to place the huge lintels atop the stone posts is staggering. But these hard rocks were not only hoisted; they were first smoothed with hammers. Incredibly, some of the rocks in Stonehenge came from quarries which were as far away as 130 miles! But the actual distance that the stones were carried was nearly twice that figure. Archaeologists have shown that the Bluestones must have been transported from the Prescelly Mountains in Wales, and that the simplest route must have covered at least 240 miles over land and water. Even with the use of rafts and rollers, this is a mind-boggling feat.

The heavier Sarsen stones were apparently brought from Marlborough Downs, about 20 miles distant from Stonehenge. This job would have required the work of 800 men.

The construction of the inner rings of Stonehenge is thought to have taken about seven years. The entire structure required an estimated total of 1.5 million man-hours of labor!

Why these prehistoric men worked so hard to construct this curious monolith remains a mystery. For centuries, it was thought that the structure was used as a pagan temple, for cremated bones were found in the Aubrey holes. But early in this century, the proposition was advanced that Stonehenge was constructed as a sort of seasonal clock, its main axis pointing directly toward the rising sun on midsummer day (June 24).

Each of the five post-and-lintel trilithons in the horseshoe frames the position of the sunset or sunrise on a key day. The Stonehenge structure could have been used by the ancient builders as a primitive alarm clock, advising when to plant and when to harvest. Calculations have shown that the risings and settings of the sun, as seen through the openings in the trilithons, are remarkably precise.

The Alhambra—the fairy-tale fortress

INCREDIBLE STRUCTURES

An invitation to choose the most beautifully ornamented building in the world will naturally result in some difference of opinion, but a structure that is sure to top many lists is the exquisite

citadel of the Alhambra, in Granada, Spain. This massive, sumptuously adorned complex of halls, towers, palaces, and courts is perhaps the finest example of the Moorish architecture that once dominated much of North Africa and Spain.

Resting atop a 35-acre plateau overlooking the historic city of Granada, the Alhambra was for many years the home of the Moorish kings, serving as palace, fortress, and administrative headquarters for their Spanish empire. Built chiefly between 1230 and 1354, the Alhambra remained the last bulwark of Islam in Europe until Granada fell to the Spaniards in 1492.

During its tempestuous history the Alhambra has survived many calamities. The Spaniards destroyed much of the citadel when they recaptured Granada. In 1812, the towers of the fortress were blown up by Napoleon's troops, and in 1821, an earthquake heavily damaged the complex. Extensive restoration was undertaken after 1828, and today the 700-year-old Alhambra has regained much of its age-old charm.

The hilltop citadel is so extensive that it would take a visitor more than an hour to walk around the surrounding walls. (It is to these red-brick walls that the citadel owes its name, for "Alhambra" means "the Red" in Arabic.) But it is not size that earns the Alhambra its fame.

The older Moorish section of the complex—the Alcazaba—is a magnificent work of interior design and sculpture. Here, intricate carvings in marble, alabaster, plaster, and glazed tile adorn the walls and ceilings. Palm-like marble pillars and stalactite vaultings form shady arcades; rich mosaics decorate the halls; delicate fountains embellish the many sun-bathed courtyards.

The Palace of the Kings is perhaps the most elegant of the many buildings in the Alhambra. In the center of this palace is the famed Court of Lions, with its alabaster basin supported by white marble beasts. Nearby, in the Hall of Ambassadors —which boasts a 75-foot high dome—water spouts through the mouths of yawning lions. In the Court of Myrtles, a 10,000 square-foot pond glistens with the reflection of nearby Myrtle trees, while underneath glimmering goldfish swim in the clear, still waters.

Through the years, the Alhambra has given birth to many legends, and ghosts are said to roam its quiet halls and courts. These spirits, according to legend, are the souls of the Alhambra's Moorish and Spanish residents, who would not forsake an earthly habitation of such heavenly beauty.

The Simplon Tunnel is over 12 miles long

The Simplon railway tunnel, cut through the Alps between Switzerland and Italy, is the longest tunnel of any kind in the world. From portal to portal, the Simplon measures 12¼ miles—very nearly the length of Manhattan Island!

The Simplon actually consists of two tunnels, each containing one railroad track. The two parallel tubes lie 56 feet apart and are connected by smaller tunnels which can serve as escape routes in the event one tunnel is flooded or caved in.

Construction of the first tunnel began in 1898. One team of workers began on the Swiss side of Mount Leone, the second team on the Italian side; both teams blasted their way toward the center of the mountain.

A project of this size naturally presented a host of problems no one could have expected. As the two construction teams bore inward through the mountain, temperatures inside the tunnel soared, rising to as high as 130° F. near the center! At that temperature, a worker could sweat away his body's salt in a very short time.

Work was halted while a water-spray system was installed to cool the tunnel walls. Then, workers on the Italian side struck a spring which poured 1,500 gallons of cold water into the tunnel every minute.

INCREDIBLE STRUCTURES

The water-spray system—thus superceded by nature—was deactivated.

It was not long before workers struck a second spring, this one sending 2,000 gallons of boiling water into the tunnel each minute! Pumps had to be rigged to clear the tunnel before work could resume.

The use of explosives and the threat of falling rock presented ever-present dangers to the workers. In all, 60 men lost their lives during the construction of the two tunnels. (This was considered a surprisingly low figure, in view of the 800 deaths incurred during the construction of the St. Gotthard tunnel in Switzerland in the 1880's.)

When the two teams met at the center of the mountain, eight years after construction began, they found their tunnels only inches apart. The discrepancy was easily corrected, and the Simplon took its place as the longest of all the world's tunnels—that is, until 1922, when the second Simplon tunnel was completed, 66 feet longer!

The Eiffel Tower is the tallest structure in Europe

In the late 1880's, the city of Paris was swept by a storm of protest over a planned tower for the upcoming International Exposition. The proposed structure — intended to symbolize the glory of France — was decried by artists, writers, and officials alike as a useless monstrosity, an affront to the history of great monuments in its use of the

INCREDIBLE STRUCTURES

then despised material, steel. Yet, despite these protests, the pressures of time, and a limited knowledge of the necessary structural techniques, Alexander Gustave Eiffel pressed on with his design. Almost 90 years later, this structure — the Eiffel Tower — is the highest (excluding TV towers), grandest, and most famous tower in the world, and the symbol of a nation.

The tallest structure in Europe (again, excluding TV towers, navigational masts, and chimneys), the Eiffel Tower rises 1,052 feet over a long landscaped promenade near the River Seine. At its base, four huge masonry pillars anchor four steel columns, which join at a height of 620 feet to form one slender spire. A football field could be placed between these pillars with room to spare.

Construction of the massive tower was completed in only 17 months in time for the International Exposition of 1889. The design, a combination of the arch and the obelisk, called for the use of 12,000 component parts weighing a total of 7,500 tons. Despite the lack of sophisticated safety measures, there were no fatal accidents during the entire period of construction. Equally surprising, the total cost of over $1 million was recovered from sightseers within *one year!*

Today, the Eiffel Tower is one of the most popular tourist attractions in the world. Elevators rise aslant along the columns to observation decks at three levels, which provide a vista of Paris and the surrounding countryside up to 90 miles away!

Although the tower was decried as useless from its very inception, from 1925 to 1936 it supported the largest advertising sign ever erected. This electric "Citroen" sign, consisting of 250,000 lamps, was visible from as far as 24 miles away. Today, the Eiffel functions as a radio and TV broadcast tower, a meteorological station, and of course, as France's greatest tourist attraction.

The Great Wall of China took 1,700 years to complete

It has been said that of all the man-made structures on the face of the earth, the only one that could conceivably be visible from the moon is the Great Wall of China. Wending its way over more than one-twentieth of the earth's circumference, the Wall is an unparalleled feat of engineering and

human determination. In size, materials, and human labor, it is the largest construction project ever undertaken by man. Enough stone was used in the entire project to build an eight-foot wall girdling the globe at the equator!

From its eastern end at Shanhaikuan on the Yellow Sea to its western end at Chaiyukuan in the Gobi Desert, the Wall stretches over mountains, deserts, and plains a distance of 1,500 miles. If the Wall were picked up and moved to the United States, it would stretch from New York to Topeka,

Kansas! But with its numerous twists and turns, the Great Wall is actually 1,700 miles long—and including all its peripheral extensions and off-shoots, the length is 2,500 miles. More than 24,000 gates and towers dot the Wall over its serpentine course.

INCREDIBLE STRUCTURES

In the eastern regions of China, the Wall is built of stone faced with brick, to an average height of 25 feet. Here the wall is generally 20 to 30 feet wide at the base, tapering to 15 feet at the top. Most portions of the eastern Wall are wide enough to permit six horsemen to ride abreast along the top.

In the west, however, the Wall is constructed largely of earth faced with stone, or simply of earth piled into mounds. Today this section of the Wall has fallen into ruin, and at points is almost obscured by drifting sands.

As the Wall stands today, it is an amalgamation of many walls built over a period of 1,700 years—making the Wall the longest continuous construction project in human history. The first emperor of China, Ch'in Shih Hwang-ti, from whose name the word *China* is derived, began building the Wall in the 3rd century B.C. Large portions of the eastern Wall were constructed during his 11-year reign. From all over the newly unified China, laborers were conscripted for the project; many died during the construction. The wall was continually augmented and improved over the centuries, with the major work being done during the Ming Dynasty (1368-1644).

Why such a gargantuan project was undertaken is not known for certain. Originally, the Wall was thought to have been built to provide a defense against Mongolian tribesmen to the north, but authorities point out that the height and the extent of the Wall made it undefendable against any army determined to invade China. Indeed, armies were successful in breaching the Wall many times throughout Chinese history. The Wall itself may have been a prime motive for some of the Mongolian invasions, for by enclosing many of the water sources in the outlying regions, the Chinese made it necessary for the tribesmen on the Mongolian plains to cross the Wall in search of water.

Some authorities maintain that the Wall was constructed solely to define the limits of Chinese sovereignity. (The Chinese penchant for walls is well demonstrated by the fortifications built around all old Chinese cities.) All land within the Wall was considered China; everything beyond, the wilderness. In fact, the Wall did serve for hundreds of years as the boundary between the Orient and the Occident.

Others maintain that the Wall was undertaken to provide employment for the Chinese masses in times of hardship and unrest. Whatever the reason for its construction, the Wall stands as one of the most incredible, and certainly the largest, construction feats ever accomplished by man.

St. Peter's is the largest Christian church in the world

For sheer size, the church of St. Peter's in Rome is an extraordinary structure, easily outstripping such great cathedrals as Rheims, Chartres, and Notre Dame de Paris. Yet St. Peter's is also one of the world's most renowned works of architecture, and boasts painting and sculpture by the greatest artists of the Renaissance. The crowning glory of an age, St. Peter's remains today the center of the Roman Catholic Church.

Even the site upon which St. Peter's rests is historically significant. It was here, on the left bank of the Tiber, that the Roman Emperor Nero built a large amphitheater to house his gory spectacles, in which thousands of Christians died for the viewing pleasure of the Roman masses. Among these victims was St. Peter, the Church's first pope, who was crucified and buried in a mass grave outside the amphitheater.

In the 4th century, the Roman Emperor Constantine the Great—the first Christian emperor—built a small church in place of the old amphitheater. The altar of this church was placed directly over the supposed site of Peter's grave. Here many popes and emperors—among them Charlemagne—were crowned.

By the 15th century, Constantine's church was

crumbling, and portions were rebuilt by Pope Nicholas V. Then, in 1506, at the height of the Renaissance, Pope Julian II decided to construct a new church on the site, a grand church worthy of the most powerful institution in the world.

Julian's plan called for a church large enough to hold 80,000 people—at that time, the entire population of Rome! A monumental design by the architect Bramante was accepted, and the greatest construction project in the Church's history was underway.

Bramante's church was so large and elaborate that 12 architects spent most of their lives working on the project. Raphael was in charge of construction for a time. Michelangelo supervised the building of the immense dome, but he—like most of the artists who worked on the project—never lived to see the church completed.

It wasn't until 1626, 120 years after construction began, that St. Peter's was dedicated. And it was another 40 years before the vast piazza and colonnades in front of the church were finished. In memory of the thousands of Christians who had died in Nero's arena, an Egyptian obelisk that had stood in the center of the amphitheater was placed in the center of this piazza.

The main section of St. Peter's is 700 feet long and 450 feet wide, enclosing an area of over seven acres. Most of the world's cathedrals could fit

INCREDIBLE STRUCTURES

inside without difficulty. Imagine a church this large topped by a roof as high as a 15-story building!

Within the church are 44 altars, the largest being a huge work of bronze upon which only the Pope himself may conduct mass. Three-hundred-ninety statues—most of them quite large—adorn the interior and exterior. But the most extraordinary feature of the church is the massive dome designed by Michelangelo. For hundreds of years, the dome of St. Peter's stood as the world's largest. It is high enough to enclose the Capitol Building in Washington, D.C.—with 65 feet to spare!

Today, St. Peter's is only a part of the Vatican complex, with its numerous chapels, palaces, and gardens. But St. Peter's remains the greatest structure of the Vatican and the greatest church in the world.

The Taj Mahal—the most beautiful building in the world

On the bank of a placid river deep in the heartland of India rests the building many people consider the most beautiful in the world—the Taj Mahal. Although in size and grandeur it is the equal of any palace or temple, the Taj Mahal was built as neither a royal residence nor a place of worship. It is the tomb of one woman, a monument to love.

About 300 years ago, Shah Jehan ruled a vast Moslem empire in India. The Emperor's wealth was enormous, his coffers filled with gold and precious jewels. His capital, Agra, was one of the most magnificent cities in the East, resplendent with marble palaces and fragrant gardens. Yet of all his treasures, the Emperor most prized his wife, Mumtaz Mahal.

When she died in 1629, the grief-stricken monarch decided to build a monument of unmatched beauty and splendor to serve as a resting place for his beloved. From all over the world, Jehan summoned architects, sculptors, and jewelers to his marble city. From Persia came boatloads of silver; from Arabia, pearls by the thousands. For 18 years, 20,000 men worked on the glorious tomb. When it was completed in 1648, the mausoleum—called the Taj Mahal—was the most beautiful building in one of the most beautiful of cities.

In size alone, the Taj Mahal is a wondrous architectural achievement. The octagonal building extends for 186 feet on its longest side, and rests on a vast marble platform 313 feet square. In each corner of the platform stands a white marble minaret, 138 feet tall. The walls of the mausoleum are 70 feet high and are topped by a massive

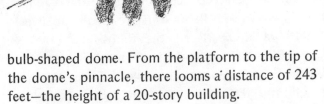

bulb-shaped dome. From the platform to the tip of the dome's pinnacle, there looms a distance of 243 feet—the height of a 20-story building.

But the true splendor of the Taj Mahal lies not in its size, but in its breathtakingly sinuous forms.

The exterior is entirely of white marble, inlaid with semi-precious stones which form Arabic inscriptions, floral designs, and arabesques. The burial room itself is surrounded by a marble screen whose intricate carvings give it the appearance of lace rather than stone. And surrounding the building on three sides is an elaborate walled garden with marble pavements, fountains, and pools that reflect the Taj in all its dream-like brilliance. The tableaux of the milk-white Taj, the emerald-green gardens, and the blue of sky and water are dazzling beyond words.

INCREDIBLE STRUCTURES

Shah Jehan had planned to build another mausoleum to house his own sarcophagus. This tomb was to be an exact duplicate of the Taj, but constructed of black marble rather than white. The twin structure would sit on the opposite bank of the Jumna River, directly across from the Taj, with a silver bridge connecting the two tombs.

However, before his plans could be carried out, Jehan was dethroned by a rebellious son. After his death, the Emperor was laid to rest beside his wife, under the dome of the Taj. Perhaps it is just as well that construction of Jehan's second tomb was never attempted, for the perfect beauty of the Taj Mahal could hardly have been equalled.